Diazo Chemistry
Synthesis and Reactions

By
John Cannell Cain

ISBN 1-929148-39-9

WEXFORD COLLEGE PRESS 2003

DEDICATED

TO THE MEMORY OF

JOHANN PETER GRIESS

IN THE 50th ANNIVERSARY OF THE DISCOVERY OF
THE DIAZO-COMPOUNDS (1858-1908)

PREFACE

In this book I have endeavoured to describe our present knowledge of the Diazo-compounds and to give an account of the enormous progress made in this important branch of Organic Chemistry since Griess's epoch-making discovery just fifty years ago.

I have adopted the plan of giving full details of the simpler preparations and reactions which are being continually carried out in the laboratory, but I have made only short reference to the more involved operations such as would be undertaken by the research chemist, not only to avoid a mass of detail, but also because in such cases the original paper is invariably consulted, and with this in view full references to the literature are given.

Owing to the very large space necessarily given to a discussion of the theories of the constitution of the diazo-compounds, the practical and descriptive portions have been kept together and the theoretical part reserved until later.

Partly on account of this the word 'diazonium' is not used until the theoretical part is reached, where its meaning is explained.

I have striven to give an exact account of the long controversy between Hantzsch and Bamberger on the constitution of the diazo-compounds. The most

important contributions to the literature have hitherto
been made by Hantzsch (*Die Diazoverbindungen*, Ahrens'
Sammlung, 1902) and his former pupil, Eibner (*Zur
Geschichte der aromatischen Diazoverbindungen*, Olden-
bourg, 1903), so that the opposite view has, perhaps,
not been set forth quite so fully, although the
admirable Report to the British Association by Morgan
(*Our Present Knowledge of Aromatic Diazo-compounds*,
1902) leaves nothing to be desired.

Another difficulty in the way of presenting a clear
view of the subject is found in the many cases where
work, even supported by numerous analyses, has been
shown to be incorrect. This, taken in conjunction
with the somewhat authoritative tone of Hantzsch's
papers and the unfortunate illness of Bamberger
which necessitated cessation of work for some years,
may well be responsible for the incomplete accounts
of Diazo-chemistry which are occasionally encountered.
Finally, a new theory of the constitution of Diazo-
compounds is presented in the Appendix to this
work.

J. C. C.

London,
December, 1907.

CONTENTS

CHAPTER I

CHAPTER II

CHAPTER III

CHAPTER IV

CHAPTER V

CONTENTS

CONTENTS

CHAPTER XIII

AZO-COMPOUNDS 80

§ 1. Azoxy-compounds.   § 2. Azo-compounds.   § 3. Aminoazo-
compounds.   § 4. Hydroxyazo-compounds.   § 5. Rate of
formation of amino- and hydroxy-azo-compounds.   § 6. Constitution of the hydroxyazo-compounds.   § 7. Mixed azo-compounds.

CHAPTER XIV

METALLIC DIAZO-DERIVATIVES. DIAZO-HYDROXIDES . 96

CHAPTER XV

DIAZO-COMPOUNDS OF THE ALIPHATIC SERIES . . 103

§ 1. Preparation.   § 2. Properties of diazoacetic esters.
§ 3. Reactions of the aliphatic diazo-compounds.   § 4. Metallic
diazo-compounds of the aliphatic series.   § 5. Diazoamino-compounds of the aliphatic series.

CHAPTER XVI

CONSTITUTION OF THE DIAZO-COMPOUNDS . . . 112

§ 1. Constitution of the diazo-salts according to Griess.
§ 2. Constitution of diazo-compounds according to Kekulé.
§ 3. Constitution of the diazo-salts according to Blomstrand.
§ 4. Constitution of diazobenzene hydroxide to 1894.

CHAPTER XVII

CONSTITUTION OF THE DIAZO-COMPOUNDS (*continued*) . 123

§ 1. Constitution of the diazo-compounds according to Hantzsch.

CHAPTER XVIII

CONSTITUTION OF THE DIAZO-SALTS AFTER 1894 . . 132

§ 1. Constitution of the diazo-compounds according to
Bamberger.   § 2. Relation between diazonium compounds
and normal or *syn*-diazo-compounds.   § 3. Double salts of
diazonium halides and metallic salts.   § 4. Diazonium halides
and *syn*-diazo-halides.   § 5. Diazonium perhalides.   § 6. Relation between *syn*- and *anti*-compounds.   § 7. The isomeric
diazo-sulphonates and diazo-cyanides.   § 8. Constitution of the
metallic diazo-oxides.   § 9. Diazo-ethers.   § 10. Diazo-
anhydrides.   § 11. Diazo-hydroxides.   § 12. Condition of the
non-ionized diazonium hydroxide.   § 13. Constitution of *iso*
(*anti*) diazo-hydroxides.

ABBREVIATIONS

ABBREVIATED TITLE.	JOURNAL.
Amer. Chem. J.	American Chemical Journal.
Annalen	Justus Liebig's Annalen der Chemie.
Arch. Pharm.	Archiv der Pharmazie.
Atti R. Accad. Lincei	Atti della Reale Accademia dei Lincei.
Ber.	Berichte der Deutschen chemischen Gesellschaft.
Bull. Acad. Sci. Cracow	Bulletin international de l'Académie des Sciences de Cracovie.
Bull. Soc. chim.	Bulletin de la Société chimique de France.
Bull. Soc. ind. Mulhouse	Bulletin de la Société industrielle de Mulhouse.
Chem. Weekblad	Chemisch Weekblad.
Chem. Zeit.	Chemiker Zeitung.
Compt. rend.	Comptes rendus hebdomadaires des Séances de l'Académie des Sciences.
Gazzetta	Gazzetta chimica italiana.
Jahresber.	Jahresbericht über die Fortschritte der Chemie.
J. pr. Chem.	Journal für praktische Chemie.
J. Russ. Phys. Chem. Soc.	Journal of the Physical and Chemical Society of Russia.
J. Soc. Chem. Ind.	Journal of the Society of Chemical Industry.
J. Soc. Dyers	Journal of the Society of Dyers and Colourists.
Journ. Chem. Soc.	Journal of the Chemical Society.
Phil. Mag.	Philosophical Magazine (The London, Edinburgh and Dublin).
Phil. Trans.	Philosophical Transactions of the Royal Society of London.
Proc.	Proceedings of the Chemical Society.
Proc. Roy. Soc.	Proceedings of the Royal Society.
Trans.	Transactions of the Chemical Society.
Zeitsch. angew. Chem.	Zeitschrift für angewandte Chemie.
Zeitsch. Elektrochem.	Zeitschrift für Elektrochemie.
Zeitsch. Farb.-Ind.	Zeitschrift für Farben-Industrie.
Zeitsch. physikal. Chem.	Zeitschrift für physikalische Chemie.
Zeitsch. f. Chem.	Zeitschrift für Chemie.
D.R-P.	Deutsches Reichs-Patent.
E.P.	English Patent.
F.P.	French Patent.

THE CHEMISTRY OF THE DIAZO-COMPOUNDS

CHAPTER I

INTRODUCTION

THE diazo-compounds were discovered in 1858 by Johann Peter Griess,* who obtained them by treating aromatic amino-compounds with nitrous acid. Piria had already found, in 1849, that asparagine or aminosuccinamic acid is converted into malic acid by the action of nitrous acid, the amino-group, NH_2, being substituted by the hydroxyl group, OH, thus—

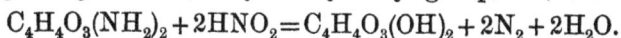

$$C_4H_4O_3(NH_2)_2 + 2HNO_2 = C_4H_4O_3(OH)_2 + 2N_2 + 2H_2O.$$

In the aromatic series, also, Hunt in the same year showed that aniline, by the same method, was converted into phenol. Then Gerland in 1853 † prepared hydroxybenzoic acid from aminobenzoic acid and also observed the formation of a red intermediate product, the quantity of which was found to increase by working with cold dilute solutions. Gerland was unable to decide as to the constitution of this substance owing to wide variations in the analytical figures. The further investigation of this was suggested to Griess by Kolbe,‡ with the result that diazoaminobenzoic acid was isolated. Griess then extended his work so successfully that he discovered the existence of an entirely new class of substances, to which the name 'diazo' was given.§ An account of this discovery, given by Griess himself, will be of interest.||

* *Annalen*, 106, 123.   † *Ibid.*, 1854, 91, 185.

‡ Obituary notices of Griess, *Ber.*, 1891, 24, 1007.

§ Griess says : 'I have come to the conclusion that the two atoms (or the molecule) of nitrogen, N_2, they contain, must be considered as equivalent to two atoms of hydrogen, and it is in accordance with this view that the names of the new compounds have been framed.' (*Phil. Trans.*, 1864, 154, 668.)

|| Roscoe and Schorlemmer, *Treatise on Chemistry*, vol. iii, part 3, 311. Griess, private communication to Watson Smith in 1887 ; compare also Watson Smith, *J. Soc. Chem. Ind.*, 1907, 26, 134.

B

'Dr. Gerland, when working in the laboratory of Prof. Kolbe, in Marburg, investigated the action of nitrous acid on amidobenzoic acid at the request of Kolbe. Thus oxybenzoic acid was prepared, indicating a chemical change then considered of much importance. In like manner I investigated a means of converting picramic acid (amidodinitrophenylic acid) into the oxydinitrophenylic acid, $C_6H_2(NO_2)_2(OH)_2$, but I obtained instead of the latter a compound possessed of such striking and peculiar properties that I at once concluded it must belong to a completely new class of compounds. Analysis soon showed me that this peculiar compound had the composition $C_6H_2(NO_2)_2N_2O$. Naturally I soon submitted many other amido-compounds in like manner to the action of nitrous acid, and obtained thus, in almost every case, the corresponding diazo-compound. But the circumstance to which I was indebted for my success in obtaining the diazo-compounds was that of the treatment of the amido-compounds with nitrous acid in the cold, whereas in the earlier experiments of Hunt and Gerland a higher temperature was always attained, and consequently no diazo-compounds could exist. Having obtained these diazo-compounds, I then tried their action on all possible substances, among which of course are the numerous class of amido-compounds. I found that the diazo-compounds combine directly with these, forming frequently brilliantly coloured substances which dye animal fibres directly. The first colouring matter thus prepared by me, which I obtained in the years 1861–2, was the benzeneazo-α-naphthylamine.* It was first prepared on the large scale, to the best of my recollection, in the years 1865–6 by Caro, who was then chemist in the works of Messrs. Roberts, Dale & Co., of Manchester. I first recommended the oxyazobenzene obtained by me for use as a colouring matter in 1866.'†

Griess continued his researches on diazo- and azo-compounds ‡ during his three years' residence in London as

---

* *Phil. Trans.*, 1864, **154**, 679.
† *Annalen*, **137**, 88.
‡ Griess's first short preliminary announcement was published in *Annalen*, 1858, **106**, 123 ; see also *Proc. Roy. Soc.*, 1859, **9**, 594 ; *Phil. Mag.* 1859 [iv], **17**, 370 ; *Compt. rend.*, 1859, **49**, 77. The full paper appeared in *Annalen*, 1860, **113**, 201. See also *Annalen*, 1860, **113**, 337 ;

Hofmann's assistant, and also afterwards while with Messrs. Allsopp, Burton-on-Trent. Here, although busily occupied in the vast brewery, Griess found time in which to prepare a large number of new diazo-compounds, and these were then handed over to his friend, Dr. R. Schmitt, at Dresden for analysis. Hempel's account of this is interesting : ' Regelmässig kamen von Burton an den Ufern des Trent die von Griess dargestellten neuen Körper in kleinen Packeten, um in Dresden an der Elbe analysirt zu werden. Per Fracht kamen dann wohl gleichzeitig als willkommene Beilage Fässer von Allsopp's berühmtem Pale Ale in ausgesuchtester Qualität.' (The new compounds prepared by Griess were regularly sent in small packets from Burton-on-Trent to Dresden to be analysed. At the same time a welcome accompaniment took the form of barrels of Allsopp's finest Pale Ale). Griess's brilliant investigations extended to the preparation of a very large number of diazo-compounds ; further, he discovered most of their reactions with other reagents and laid the foundation of the immense edifice of azo-dyestuffs which has since been, and is still being, erected.

After having described the diazoamino-derivatives of aminobenzoic acid,[*] aminotoluic acid, and aminoanisic acid, and their reactions, Griess then obtained diazoaminobenzene [†] and the diazobenzene salts.[‡] Of these the nitrate, the easiest to prepare, was used as the starting-point in the preparation of the crystallized sulphate, the platinichloride, the aurichloride, and, some years later,[§] the ferricyanide, the nitroprussiate, and the tin chloride double salt. With the object of preparing the bromide, Griess treated diazoaminobenzene with bromine in ethereal solution and obtained the perbromide $C_6H_5N_2Br_3$.

1861, **117**, 1 : **120**, 125 ; 1861, Suppl. I, 100 ; 1862, **121**, 257 ; 1866, **137**, 39 ; *Proc. Roy. Soc.*, 1860, **10**, 309, 591 ; 1862, **11**, 263 ; 1863, **12**, 418 ; 1864, **13**, 375. A long paper, including most of the earlier work, is in *Phil. Trans.*, 1864, **154**, 667 ; and accounts were also published in *Journ. Chem. Soc.*, 1865, **3**, 268, 298 ; 1866, 4, 57 ; 1867, **5**, 36. For the later work see *Ber.*, 1869, 2, 369 ; 1874, **7**, 1618 ; 1876, **9**, 132, 627, 1653 ; 1878, **11**, 624 ; 1879, **12**, 2119 ; 1881, **14**, 2032 ; 1882, **15**, 2183 ; 1883, **16**, 2028 ; 1884, **17**, 338.
* *Annalen*, 1861, **117**, 1. † *Ibid.*, 1862, **121**, 257.
‡ *Ibid.*, 1866, **137**, 39.
§ *Ber.*, 1879, **12**, 2119 ; 1885, **18**, 965.

On the addition of ammonia to this compound, the whole of the bromine was removed and a substance containing three atoms of nitrogen was isolated. This was called diazobenzeneimide and possessed the formula $C_6H_5N_3$. Griess also studied the formation of metallic diazo-derivatives; thus, by the action of a concentrated potassium hydroxide solution on a strong solution of diazobenzene nitrate, a substance containing potassium was obtained, and by mixing a solution of this with a silver solution, a substance containing silver was precipitated. The formulae of these metallic derivatives were considered by Griess to be $C_6H_5N_2.OK$ and $C_6H_5N_2.OAg$ respectively (see, however, p. 96). By treating a solution of the potassium compound with acetic acid, a viscous yellow oil was obtained which Griess looked upon as free diazobenzene. With mineral acids it yielded the corresponding salts.

We now pass on to consider briefly the various reactions which the diazo-salts, in Griess's hands, were found to undergo. By boiling with water, phenols were obtained; in the case of the nitrate, nitrophenol was formed by the interaction of phenol and the liberated nitric acid. When alcohol was substituted for water benzene was formed, whilst the alcohol was reduced to aldehyde. By the action of hydriodic acid in the cold, iodobenzene was obtained, chlorobenzene by distilling the dry platinichloride with soda, and bromobenzene in the same way from the platinibromide, and also by boiling the perbromide with alcohol.

By the action of phenols and amines on the diazo-compounds Griess discovered that highly-coloured condensation products were formed. These were the azo-dyestuffs, some of which were immediately prepared on the large scale; the reaction itself giving the key to an industry which has since attained an enormous importance.

The next important discovery to be noted is that of the first diazo-compound belonging to the aliphatic series in 1883. Curtius prepared the ethyl ester of diazoacetic acid, proceeding from this to a series of brilliant researches on fatty diazo-compounds, culminating in his discovery of azoimide. In 1894 von Pechmann isolated the simplest member of the series, namely, diazomethane, since which time a large number of derivatives have been obtained.

The remarkable influence of small amounts of copper salts on the reactions which the diazo-compounds undergo was discovered in 1884 by Sandmeyer, whose name is associated with this decomposition, and the substitution of finely divided copper for its salts was introduced by Gattermann in 1890.

In the latter year Meldola discovered that the presence of the diazo-group has, in certain cases, a remarkable effect on the stability of a nitro-group present in the same benzene ring, whereby this group is very readily eliminated.

This transformation of acidic groups under the influence of the diazo-group has been made the subject of comprehensive researches by Meldola and his colleagues, as well as by Bamberger, Orton, and Hantzsch, and the latter chemist showed in 1896 that in some cases the acidic group attached to the diazo-nitrogen could change places with a halogen atom in the benzene ring.

In 1894 an important investigation carried out by Schraube and Schmidt, whereby the existence of two isomeric metallic salts of diazobenzene was indicated, led to a thorough examination of the metallic diazo-derivatives, and gave rise to a prolonged discussion of the constitution of the whole class of diazo-compounds. Among the many discoveries which were made in 1895 is specially to be recorded Bamberger's isolation of the diazoic acids by the oxidation of the diazo-salts.

The definite proof by Andresen, in that year, that light acted on diazo-compounds with the production of the corresponding phenols was followed by the remarkable observation of Orton in 1906, that quantitative yields of phenols were obtained from certain diazo-salts which gave practically no hydroxy-derivative when heated with water or acids.

CHAPTER II

PREPARATION OF THE DIAZO-COMPOUNDS.*

§ 1. Preparation of dry diazo-salts.—In preparing the diazo-compounds Griess used, as a source of nitrous acid, the gases evolved by warming a mixture of nitric acid and arsenic trioxide. These gases were passed into either an alcoholic solution of the amine or an aqueous paste of an amino-salt, the experiment being carried out in the cold, when the resulting diazo-compound separated or was precipitated by the addition of alcohol and ether.

In the case of diazobenzene nitrate, used by Griess as the starting-point for the preparation of other diazo-compounds, the nitrous gases were passed into a well-cooled paste of aniline nitrate and water until aniline ceased to be liberated on adding potassium hydroxide to a small test portion. The solution was then filtered and alcohol and ether added to precipitate the diazobenzene nitrate, which separated in white needles.

A considerable improvement on this method consists in the use of a solution of sodium nitrite as a source of nitrous acid.† By this process the calculated quantity of nitrous acid may be used, and this is so convenient that sodium nitrite is almost entirely employed at the present day in the preparation of diazo-compounds.

The use of an aqueous solution of sodium nitrite is, however, not very suitable for the preparation of those dry diazo-compounds which are very soluble in water, and in order to avoid the presence of the latter Knoevenagel ‡ used amyl nitrite in alcoholic solution, a method which had been em-

* The description of the *metallic* diazo-compounds is reserved until a later chapter (see p. 96).
† Martius, *J. pr. Chem.*, 1866, **98**, 94.
‡ *Ber.*, 1899, **23**, 2995.

ployed by V. Meyer and Ambühl in the preparation of diazo-aminobenzene.* This was a great improvement on existing methods, and a large number of dry diazo-salts have been prepared in this way. Pure products are, however, only obtained in the absence of free mineral acid.† The reason of this is that in presence of excess of mineral acid additive compounds of the diazo-chloride with hydrochloric acid are formed.

The hydrochlorides are therefore prepared by passing dry hydrogen chloride into a solution of the amine in absolute alcohol or ether and heating the product at 40–50° until the last traces of acid have been removed. The dry salt is then dissolved in alcohol and the theoretical quantity of amyl nitrite added at the ordinary temperature. On precipitating with ether the diazo-chloride is obtained in the pure state. This is a somewhat tedious process, and it has been found ‡ that the reaction proceeds even more satisfactorily in the presence of glacial acetic acid, thus avoiding the necessity of preparing the dry aminic hydrochloride.

In order to prepare diazobenzene chloride the experiment is carried out as follows: Fifty grams of aniline hydrochloride are dissolved or suspended in about three times the quantity of glacial acetic acid and the mixture stirred by a turbine. A little more than the theoretical quantity of amyl nitrite is now added, care being taken that the temperature does not exceed 10°. Any undissolved aniline salt disappears quickly, and the diazotization is complete as soon as a small portion withdrawn and treated with sodium acetate no longer gives a yellow coloration. On adding ether a thick crystalline precipitate of diazobenzene chloride is obtained, which is filtered, washed with ether, and dried in a desiccator. The yield is 53 grams. The sulphate is obtained in a similar way; in this case aniline sulphate is diazotized in presence of the calculated quantity of sulphuric acid. The separation of the diazo-sulphate is effected by first adding a little alcohol to the mixture with acetic acid and then precipitating with ether. The halogenated diazo-chlorides are prepared in the

* *Annalen*, 1889, **251**, 56.　　　† Hirsch, *Ber.*, 1897, **30**, 1148.
‡ Hantzsch and Jochem, *Ber.*, 1901, **34**, 3337.

8 CHEMISTRY OF THE DIAZO-COMPOUNDS

same way, but care must be taken to use the hydrochlorides obtained according to Hirsch's method, avoiding the presence of mineral acid.

The preparation of diazo-salts which are sparingly soluble in water can, of course, be carried out in aqueous solution. In some cases the diazo-compound is precipitated on addition of sodium nitrite to the acid solution of the amine, whilst in others the insoluble diazo-compound is precipitated on adding the salt of a different acid.

Many instances of the former case occur among the amino-sulphonic acids. Thus p-diazobenzenesulphonic acid is easily obtained * by dissolving sulphanilic acid in dilute aqueous sodium hydroxide, acidifying with hydrochloric acid, and adding the calculated quantity of sodium nitrite, previously dissolved in a small quantity of water. The temperature should be about 5°. As soon as the whole of the nitrite has been added the diazo-compound, in the form of the anhydride $C_6H_4O_3N_2S$, separates in fine white needles, which may be filtered, but should not be dried, as they are extremely explosive (see p. 28). The diazo-compound derived from a-naphthylamine-4-sulphonic acid (naphthionic acid) is obtained in a similar manner. The naphthionic acid is dissolved in alkali and reprecipitated by the addition of mineral acid. On adding the nitrite solution, the white insoluble naphthionic acid is gradually converted into the yellow insoluble diazo-compound.

Instances of the second method are numerous; thus on adding a solution of sodium picrate to a solution of diazo-benzene nitrate a precipitate of the insoluble diazobenzene picrate is obtained.† A very stable diazo-picrate has also been prepared from p-aminobenzanilide in the same way,‡ and the picrate of diazophenylindole, $C_{20}H_{12}O_7N_6$, as also the picrate of diazomethylindole, $C_{15}H_{10}O_7N_6$, can be crystallized from alcohol.§ The chromate was prepared by Griess and Caro,‖ who diazotized aniline nitrate by means of a solution

* Schmitt, *Annalen*, 1859, **112**, 118 ; 1861, **120**, 144.
† Baeyer and Jaeger, *Ber.*, 1875, **8**, 894.
‡ Morgan and Wootton, *Proc.*, 1906, **22**, 23.
§ Castellana and d'Angelo, *Atti R. Accad. Lincei*, 1905 [v], **14**, ii. 145.
‖ *Jahresber.*, 1867, 915.

of calcium nitrite, and then added an equivalent of potassium dichromate and hydrochloric acid, when a precipitate was obtained. They suggested the use of this chromate as an explosive.*

Diazo-chromates are now usually prepared † by precipitating a diazo-solution with sodium dichromate.

Insoluble diazo-thiosulphates, hydroferricyanides, and tungstates have been also prepared in a similar manner.‡

An interesting diazo-carbonate is obtained § by pouring the diazo-chloride derived from benzoyl-p-phenylenediamine (p-aminobenzanilide) into cold aqueous sodium carbonate, when a yellow precipitate results which has the formula $C_6H_5 . CO.NH.C_6H_4 . N_2 . HCO_3$.

By adding sodium acetate to the diazo-chloride, and then treating with excess of sodium nitrite, a crystalline yellow diazo-nitrite is obtained.

Certain diazo-salts of hydrazoic acid have also been prepared by treating an ethereal solution of the diazohydroxide with ethyl azoiminocarboxylate, $N_3 . CO_2Et$. These salts have the composition $Ar . N_2 . N_3$, and are extremely unstable.‖

Diazo-fluorides containing one molecule of hydrofluoric acid are obtained by diazotizing the amine with amyl nitrite in presence of hydrofluoric acid,¶ and diazo-perchlorates are produced by diazotizing amines in presence of perchloric acid. These perchlorates are extremely explosive.**

§ 2. **Other methods of preparation.**—A modification of Griess's method of diazotizing was used by his co-worker Schmitt,†† who saturated absolute alcohol with nitrous fumes and poured this over aminophenol hydrochloride. On adding ether the diazophenol was precipitated.

* *Bull. Soc. chim.*, 1867 [ii], **7**, 270 ; F. P. 73286.
† Meldola and Eynon, *Trans.*, 1905, **87**, 1 ; Castellana and d'Angelo, loc. cit.
‡ Hepburn, *J. Soc. Dyers*, 1901, **17**, 279.
§ Morgan and Micklethwait, *Trans.*, 1905, **87**, 921.
‖ Hantzsch, *Ber.*, 1903, **36**, 2056.
¶ Hantzsch and Vock, *Ber.*, 1903, **36**, 2059.
** *Ber.*, 1906, **39**, 2713, 3146.
†† *Ber.*, 1868, **1**, 67.

The direct interaction of fuming nitric acid and amines for the preparation of diazo-nitrates is hardly to be classed as a separate method, as fuming or brown nitric acid always contains nitrous fumes which diazotize the nitrate of the amine. It is interesting, however, in this connexion to note that Stenhouse * obtained the diazodinitrophenol of Griess by pouring *boiling* nitric acid on picramic acid, and several diazonitrotoluenesulphonic acids have been obtained in the dry state by diluting a solution of the corresponding toluidine-sulphonic acid in fuming nitric acid; nitration and diazotization both having been effected.† A more recent example of this is the preparation of 2-nitro-4-diazo-m-xylene-6-sulphonic acid from m-xylidinesulphonic acid—

$$(CH_3)_2 : NH_2 : SO_3H = 1:3:4:6. ‡$$

A surprising reaction is the formation of the diazo-derivative of aminophenolsulphonic acid by treating it with nitric acid and carbamide, § for one would expect the nitrous acid present to be destroyed by the carbamide and thus prevent diazotization, but recent experiments ‖ have shown that this destruction is very incomplete in the case of concentrated nitric acid. Various other nitrous derivatives have been used occasionally in the preparation of diazo-compounds; it is even stated that nitric oxide can be substituted for nitrous fumes in the preparation of diazobenzene nitrate.¶

Other substances which have been used are nitrosyl bromide ** and chloride,†† and nitrosulphonic acid.‡‡

The use of barium nitrite instead of sodium nitrite has been suggested for preparing dry diazo-salts.§§ By using the calculated quantity of sulphuric acid the whole of the mineral

* *Journ. Chem. Soc.*, 1868, **6**, 150.
† Limpricht, *Ber.*, 1874, **7**, 452.
‡ Zincke, *Annalen*, 1905, **339**, 202.
§ Bennewitz, *J. pr. Chem.*, 1874 [ii], **8**, 50.
‖ Silberrad and Smart, *J. Soc. Chem. Ind.*, 1906, **25**, 156.
¶ Ladenburg, *Ber.*, 1875, **8**, 1212.
** Koninck, *Ber.*, 1869, **2**, 122.
†† Pabst and Girard, D. R.-P. 6034 of 1878, and *Ber.*, 1879, **12**, 365 ; compare also Kastle and Keiser, *Amer. Chem. J.*, 1895, **17**, 91, who obtained a double salt, diazobenzene aniline chloride, $C_6H_5N_2Cl, C_6H_5NH_3Cl$, by treating aniline hydrochloride with nitrosyl chloride.
‡‡ Pabst and Girard, loc. cit.   §§ Witt, *Ber.*, 1903, **36**, 4388.

matter is precipitated, and, on filtering, the diazo-salt may be precipitated (with alcohol and ether) free from admixture with inorganic salts.

It has also been found possible to obtain diazo-compounds without using an amine as the starting-point; thus dry diazobenzene nitrate has been prepared * by the action of nitrous fumes on mercury diphenyl,

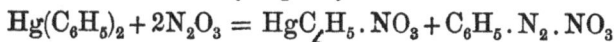

$$Hg(C_6H_5)_2 + 2N_2O_3 = HgC_6H_5 . NO_3 + C_6H_5 . N_2 . NO_3$$

and when mercury p-ditolyl is substituted for the diphenyl compound, p-diazotoluene nitrate is formed.† By treating a solution of nitrosobenzene in chloroform with nitric oxide, diazobenzene nitrate was obtained by Bamberger,

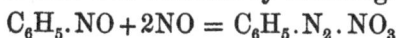

$$C_6H_5 . NO + 2NO = C_6H_5 . N_2 . NO_3$$

and the same diazo-salt has been isolated by passing nitrous fumes into an ethereal solution of nitrosophenylhydrazine.‡ Finally, by the action of alcoholic hydrochloric acid on nitroso-anilidoacetic acid, the chloride of p-diazophenylhydroxylamine, $OH.NH.C_6H_4 . N_2Cl$, is produced. §

§ 3. Diazotization of amino-phenols and -thiophenols.

Quinonediazides (Diazophenols).—The diazo-chlorides of o- and p-aminophenol are obtained by diazotizing the corresponding bases in alcoholic solution with amyl nitrite and hydrochloric acid at 0°, and precipitating with ether. ‖ The diazo-salts thus obtained are white. m-Diazophenol chloride is extremely unstable, and loses nitrogen even at 0°.

When the two former salts are dissolved in water and treated with potassium hydroxide or moist silver oxide, hydrochloric acid is split off, and the free diazophenols, or quinonediazides,¶ are formed—

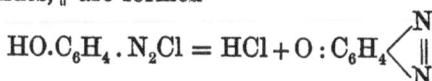

$$HO.C_6H_4 . N_2Cl = HCl + O : C_6H_4 {<} \begin{matrix} N \\ \| \\ N \end{matrix}$$

These quinonediazides are yellow and soluble in water.

* Ber., 1897, **30**, 509.   † Kunz, Ber., 1898, **31**, 1528.
‡ Rügheimer, Ber., 1900, **33**, 1718.
§ O. Fischer, Ber., 1899, **32**, 247.
‖ Schmitt, Ber., 1868, **1**, 67. Hantzsch and Davidson, Ber., 1896, **29**, 1522. See also Cameron, Amer. Chem. J., 1898, **20**, 229.
¶ For constitution see p. 126.

The substituted aminophenols are converted directly into
the quinonediazides on diazotization. The compound derived
from aminodinitrophenol (picramic acid) was the first diazo-
compound obtained by Griess.

Quinonediazides are also formed by allowing neutral solutions (or solutions containing no free mineral acid) of certain
substituted diazo-salts to stand for some time; thus 2 : 4 : 6-tri-
chlorodiazobenzene hydrogen sulphate or nitrate loses one
atom of chlorine and becomes converted into 3:5-dichloro-
o-quinonediazide,

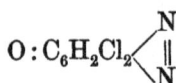

$$O:C_6H_2Cl_2\diagup\!\!\!\diagdown\begin{matrix}N\\ \|\\ N\end{matrix}$$

and many other halogen substituted anilines behave in the
same way.*

§ 4. **Thiodiazoles (Diazosulphides).**—When o-aminophenyl-
mercaptan

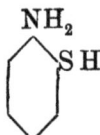

$$\begin{matrix}NH_2\\ \bigcirc S H\end{matrix}$$

is treated with nitrous acid an anhydride is obtained, as in
the case of the aminophenols, but, unlike the quinonediazides,
the o-diazosulphides are colourless, resembling in this respect
the azimides obtained by the action of nitrous acid on the
o-diamines. They generally crystallize well, have a characteristic sweetish odour, and are very feebly basic. Their
constitution is probably represented as

$$C_6H_4\diagup\!\!\!\diagdown\begin{matrix}S\\ N\end{matrix}\diagup\!\!\!\diagdown N \dagger$$

§ 5. **Preparation of diazo-salts in aqueous solution.**—From
the foregoing it will have been seen that the preparation
of solutions of diazo-salts is a comparatively simple matter,
nevertheless there are many amines, mostly substituted, which

* Orton, *Proc. Roy. Soc.*, 1903, **71**, 153 ; *Trans.*, 1903, **83**, 83, 796.
† Jacobson, *Annalen*, 1893, **277**, 209, 218, 232, 237.

either resist the action of nitrous acid or, owing to the formation of secondary products, are incapable of yielding diazo-
salts. These will be dealt with later.

In all cases the method of preparation on the large and the
small scale is the same, so that a technical recipe may be
exactly imitated in the laboratory and vice versa.

The amine to be diazotized is usually dissolved in about 10
parts of water with addition of one equivalent of hydrochloric
acid, if necessary by the aid of heat. The solution is then
cooled to 0–5° by the direct addition of ice and one and a half
to two equivalents of hydrochloric acid added. (When the
hydrochloride of the base is easily soluble in hydrochloric
acid the whole of the acid may be used in dissolving the
amine.) A solution of the calculated quantity of sodium
nitrite is now added; in most cases slowly until only a weak
reaction is obtained with starch-iodide paper (this is best
prepared from cadmium iodide and starch) after the solution
has stood for 3–4 minutes. But in certain cases, especially
where there is a great tendency towards the formation of
diazoamino-derivatives, as in the case of α-naphthylamine and
p-nitroaniline, the nitrite solution is added all at once, the
precaution being taken of adding sufficient ice to prevent the
temperature rising unduly. Occasionally the secondary reaction may be avoided by using a nitrite solution which has been
previously acidified with hydrochloric acid or by using a
larger excess of acid. The tendency towards the formation of
diazoamino-compounds increases if organic acids are used,
thus, for example, if two and a half equivalents of acetic acid
are substituted for the same equivalent quantity of hydrochloric acid, in the case of aniline only about 20 per cent. is
converted into the diazo-salt; the diazotization is complete
only by the use of eleven equivalents of acetic acid. The use
of less hydrochloric acid has a similar effect; aniline hydrochloride is only partly diazotized by sodium nitrite, but the
quantity converted increases with the concentration of the
solution, thus in solutions containing respectively 10,1, and
0·1 per cent. of aniline, about 30, 20, and 10 per cent. of the
aniline is diazotized.*

* Altschul, *J. pr. Chem.*, 1896 [ii], **54**, 508.

One of the most frequently prepared diazo-compounds is that derived from *p*-nitroaniline; indeed it is stated * that more than 1,000 tons of *p*-nitroaniline are yearly converted into the diazo-compound for the purpose of producing 'para-nitraniline red' by combination with β-naphthol on the cotton fibre.

A large number of methods of preparing this important diazo-compound have been published, † one of which (Cassella & Co.) is here quoted.

p-Nitroaniline (21 grams) is dissolved in water with addition of 42 c.c. of hydrochloric acid of 22 Bé, and the solution cooled to 5–10°. The water and ice used weigh 307 grams. A solution of 11·5 grams of technical sodium nitrite (95 per cent.) in 103·5 grams of water is now added all at once and the mixture well stirred until a clear solution is obtained. If the resulting diazo-solution is to be used for combination with β-naphthol it is first treated with a solution of 25 grams of sodium acetate dissolved in 50 grams of water. ‡

Generally speaking, amines such as aniline, the toluidines, the xylidines, *p*-aminoacetanilide, are diazotized at 0–2°. Others, as for example, α- and β-naphthylamines, the nitroanilines, and diamines, such as benzidine, tolidine, dianisidine, are converted into the diazo-compounds more suitably at about 10°. § Hydrochloric acid is most commonly employed, but sulphuric and acetic acids are also used.

There are many cases where the diazotization of an amino-compound is not effected quite so easily as is described above, and special methods have to be employed. Thus many aminoazo-compounds are insoluble in water or acids and are attacked by nitrous acid only with difficulty. At the same time the diazo-compound is often insoluble in water. Such compounds are, for example, *p*-sulphobenzeneazo-α-naphthyl-amine and *p*-acetylaminobenzeneazo-α-naphthylamine, and these are diazotized by using an excess of sodium nitrite and

* Schwalbe, *Zeitsch. Farb. Ind.*, 1905, 4, 433.
† Schwalbe, loc. cit.
‡ Compare also Schwalbe, *Zeitsch. Farb. Ind.*, 1905, 4, 433; Erban and Mebus, *Chem. Zeit.*, 1907, 31, 663, 678, 687, 1011.
§ For a detailed description of the preparation of a number of these see Cain and Thorpe, *The Synthetic Dyestuffs*, 1905, 226 et seq.

stirring for several hours, keeping the mixture ice-cold in order to avoid escape of nitrous acid. In order to prevent this escape it has been proposed to diazotize under increased pressure.* The amine is introduced into a closed vessel together with the corresponding quantity of mineral acid, and the pressure is then raised by admitting compressed air or other indifferent gas, after which the nitrite solution is added.

Difficulties have been met with in attempting the diazotization of substituted amines containing a number of acidic groups; thus V. Meyer and Stüber † found it impossible to decompose trinitroaniline

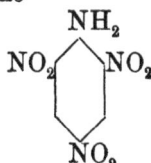

$$NH_2$$
$$NO_2 \diagup\!\!\diagdown NO_2$$
$$NO_2$$

by treatment with ethyl nitrite in alcoholic solution, and pentabromoaniline also resists diazotization ‡ unless a large excess of sulphuric acid is employed.§

This method‖ is found to be advantageous in diazotizing derivatives of aniline containing several negative groups. The base is dissolved in sulphuric acid (monohydrate), the solution cooled to $-10°$ to $-15°$, and a very concentrated solution of sodium nitrite added in excess during 1–1½ hours, the liquid being well stirred. On diluting the solution any unaltered amine is often precipitated and can be removed by filtration. Fuming, 40 per cent., hydrochloric acid may sometimes be used instead of sulphuric acid.

This method has been successfully applied in diazotizing dinitro-p-toluidine

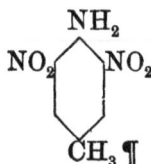

$$NH_2$$
$$NO_2 \diagup\!\!\diagdown NO_2$$
$$CH_3 \ \P$$

* Seidler, D. R-P. 143450.   † Annalen, 1873, 165, 187.
‡ Noelting, Bull. Soc. ind. Mulhouse, 1887, 57, 30.
§ Hantzsch, Ber., 1900, 33, 520.
‖ Claus and Wallbaum, J. pr. Chem., 1897 [ii], 56, 48.
¶ Claus and Beysen, Annalen, 1891, 266, 224.

Although, as Griess found, the ortho-aminophenols can be easily diazotized (see p. 11), when the corresponding compounds of the naphthalene series are similarly treated, difficulties often arise owing to the oxidizing action of the nitrous acid. This is especially applicable to the 1:2- and 2:1-amino-naphthols. In order, therefore, to obtain diazo-salts derived from these substances a number of methods have been employed with the object of avoiding this action. Thus the addition of copper or zinc salts to the solution of the amine or the use of the nitrites of zinc, nickel, mercury, &c., has been found efficacious.* For example, 12 kilos of 1-amino-2-naphthol-4-sulphonic acid are mixed with 50 litres of water and ice and a solution of 1 kilo of copper sulphate added. A solution of 3·5 kilos of sodium nitrite is now slowly run in, and after diazotization is complete the solution is filtered and the diazo-compound precipitated with hydrochloric acid. This diazo-compound may be dried and powdered.† A second example is the following : ‡ 48 kilos of the above acid are mixed and well stirred with a solution of 33 kilos of zinc sulphate in 33 litres of water containing a little zinc hydroxide. The latter is formed by the addition of about 3 kilos of ammonia to the solution. A concentrated aqueous solution of 14 kilos of sodium nitrite is then added. The reaction is completed by warming for two hours at about 40°, and the mass is then acidified with acetic acid. By filtration and crystallization, brilliant bronze needles of the diazo-compound are obtained.

Another method consists in carrying out the diazotization of these amino-compounds in presence of excess of acetic or oxalic acid. § The diazo-compounds derived from such amino-hydroxynaphthalenesulphonic acids are so stable that they can be sulphonated ‖ and nitrated. ¶

It is singular that the 2 : 3-aminonaphthols can be smoothly diazotized in the usual manner.**

§ 6. **Other methods of preparing solutions of diazo-salts.—** In addition to the methods given in §§ 1 and 2 which were used to obtain dry diazo-salts, many other ways of producing

* E. P. 10235 of 1904. See also D. R-P. 171024, 172446.
† E. P. 15025 of 1904.   ‡ F. P. 353786 of 1905.
§ D. R-P. 155083, 175593.   ‖ D. R-P. 176618, 176620.
¶ D. R-P. 164665, 176619.   ** E. P. 28107 of 1897.

these compounds in solution have been used. For example, diazobenzene chloride can be obtained by the action of zinc dust and hydrochloric acid on a solution of aniline nitrate,* thus

$$C_6H_5 . NH_2, HNO_3 + Zn + 3HCl = C_6H_5 . N_2Cl + ZnCl_2 + 3H_2O.$$

It is obvious that the action is a reducing one, the nitric acid being converted into nitrous acid by the nascent hydrogen.

A reaction similar to this is the production of diazo-salts from the nitrites of aromatic amines by treatment with a mineral acid.†

Bamberger found ‡ that when nitrosoacetanilide,

$$C_6H_5 . NAc.NO,$$

was triturated with excess of 50 per cent. potassium hydroxide, the resulting solution showed the presence of a diazo-compound. (The nature of this, existing in an alkaline solution, will be explained later.)

Certain nitroso-compounds, which contain the nitroso-group in the benzene nucleus may be directly converted into diazo-compounds by the action of three molecular proportions of nitrous acid, thus

$$R.NO + 3HNO_2 = R.N_2 . NO_3 + HNO_3 + H_2O. §$$

The method has been successfully applied to the preparation of the diazo-derivative of diphenylamine from the *p*-nitroso-compound.‖

Quinoneoxime also, when treated with nitrogen trioxide in ethereal solution, yields the corresponding diazo-salt.¶

The formation of diazo-compounds by the interaction of nitrogen peroxide and quinonedioximes ** is of much interest from a theoretical point of view (see p. 163). When, for example, thymoquinonedioxime is treated with nitrogen peroxide a nitrosodiazo-derivative is obtained.

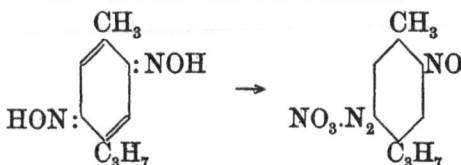

* Möhlau, D. R-P. 25146, *Ber.*, 1883, **16**, 3080.
† Wallach, *Annalen*, 1907, **353**, 322.
‡ *Ber.*, 1894, **27**, 915 ; compare E. P. 13577 of 1894.
§ O. Fischer and Hepp, *Annalen*, 1888, **243**, 282.
‖ Hantzsch, *Ber.*, 1902, **35**, 894.　　　¶ Jaeger, *Ber.*, 1875, **8**, 894.
** Oliveri-Tortorici, *Gazzetta*, 1900, **30**, i. 526.

Diazo-compounds are also obtained by the oxidation of phenylhydrazines with mercuric oxide * or acetate,† with nitrous acid in presence of a strong mineral acid,‡ and with bromine,§ and also by the action of acidic chlorides on thionylphenylhydrazone.‖

An electrolytic process for the preparation of diazo-salts has been patented by Boehringer & Sons.¶ As the method is carried out at temperatures of from 40° to 90°, under which conditions the diazo-salt would be very quickly decomposed, the latter is immediately combined with a hydroxyl compound, such as β-naphthol-3:6-disulphonic acid ('R salt'). A solution containing a mixture of aniline, sodium nitrite, and R salt is charged into a suitable cell at the platinum electrode whilst dilute sodium hydroxide surrounds the nickel cathode. On electrolyzing, the diazo-salt is formed and at once condenses with the R salt, with production of the azo-dyestuff.

A somewhat indirect method of obtaining diazo-salts was observed by Lauth,** who found that certain azo-dyestuffs were split up into quinones and diazo-compounds by treatment with an oxidizing agent, such as lead peroxide and sulphuric acid. It has been found also †† that by the action of red fuming nitric acid on azo-dyestuffs, a reaction first studied by Meldola,‡‡ oxidation and nitration takes place, and diazo-compounds, together with nitro-derivatives of the second constituent of the dyestuff, are formed.

§ 7. The action of nitrous acid on aromatic substances containing more than one amino-group.—In investigating the action of nitrous acid on diamines or triamines of the aromatic series, one would expect each amino-group to become converted into the corresponding diazo-group. Although in most cases, perhaps, this is the primary action, yet very often some secondary reaction ensues with such rapidity that no

* E. Fischer, *Annalen*, 1879, **199**, 320.
† Bamberger, *Ber.*, 1899, **32**, 1809.
‡ Altschul, *J. pr. Chem.*, 1896 [ii], **54**, 496.
§ Michaelis, *Ber.*, 1893, **26**, 2190.
‖ *Annalen*, 1892, **270**, 116.
¶ D. R.-P., 152926 of 1902; E.P. 2608 of 1904.
** *Bull. Soc. chim.*, 1891 [iii], **6**, 94.
†† O. Schmidt, *Ber.*, 1905, **38**, 3201.
‡‡ *Proc.*, 1894, **10**, 118; *Trans.*, 1889, **55**, 608; 1894, **65**, 841.

diazo-salt can be isolated by the usual means, and special methods have to be employed.

Very striking differences in behaviour are exhibited by the three phenylenediamines. When a dilute solution of sodium nitrite is added to a dilute solution of the sulphate of o-phenylenediamine, aziminobenzene is formed * according to the equation—

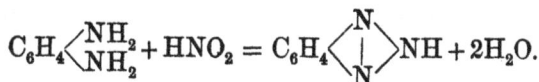

$$C_6H_4{\diagdown{NH_2}\atop\diagup^{NH_2}} + HNO_2 = C_6H_4{\diagdown{N}\atop\diagup_{N}}{\bigg\rangle}NH + 2H_2O.$$

Griess obtained this substance by acting on o-phenylenediamine hydrochloride with p-diazobenzenesulphonic acid.† It has not yet been found possible to prepare the tetrazocompound. In the case of o-tolylenediamine a similar reaction takes place, and it has been shown ‡ that the amino-group which is in the meta-position relative to the methyl group is converted into the diazo-group before internal condensation, resulting in the formation of the azimino-compound, takes place.

m-Phenylenediamine behaves very differently. When a solution of the hydrochloride is treated with sodium nitrite, the well-known dyestuff 'Bismarck Brown' is obtained. This is the hydrochloride of bisbenzeneazophenylenediamine. If the nitrite is added suddenly, a certain amount of nitroso-m-phenylenediamine is formed.§

When the reaction, however, is carried out in a different manner, both amino-groups may be diazotized, forming a tetrazo- or bisdiazo-compound. Thus Griess ‖ showed that the reaction could be successfully brought about by taking care that both the nitrite solution and hydrochloric acid are always in excess of the diamine. A two per cent. solution of m-phenylenediamine hydrochloride is prepared, and, on the other hand, a dilute solution of sodium nitrite of specific gravity 1.1. To the latter is added an equal volume of hydrochloric acid of specific gravity 1.15, and then the diamine solution added slowly, keeping the mixture well

---

* Ladenburg, *Ber.*, 1876, **9**, 221. † *Ber.*, 1882, **15**, 2195.
‡ Noelting and Abt., *Ber.*, 1887, **20**, 2999.
§ Täuber and Walder, *Ber.*, 1900, **33**, 2116. ‖ *Ber.*, 1886, **19**, 317.

stirred until the dark yellow solution of the tetrazo-compound is obtained. A later modification of this method is the following: * 80 c.c. of fuming hydrochloric acid are diluted with about 400 grams of ice and cooled with a freezing mixture. To this is added a solution of 15 grams of sodium nitrite in cold water, so that a strong solution of nitrous acid is obtained. To this solution is added quickly a cold solution of 9 grams of m-phenylenediamine hydrochloride to which 10 c.c. of strong hydrochloric acid had been added. The mixture is well stirred during the operation, and a clear yellow solution of the tetrazo-compound is obtained. Other methods consist in adding the nitrite solution to a mixture of the diamine with a large excess of hydrochloric acid,† or in pouring a mixture of the diamine and nitrite into ice-cold dilute hydrochloric acid.‡ The dry tetrazo-chloride has also been prepared.§

The tetrazo-compound derived from m-tolylenediamine is prepared similarly.‖

Substituted m-phenylenediamines, containing the substituent attached to that carbon atom which is in the ortho-position to both amino-groups are, as a rule, easily tetrazotized. Thus the m-tolylenediaminesulphonic acid of formula

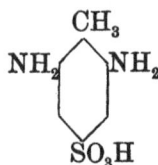

is tetrazotized without difficulty,¶ as are also such diamino-hydroxy-compounds as those of the formulae

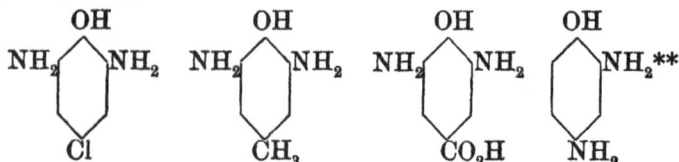

* Täuber and Walder, *Ber.*, 1897, **30**, 2901.
† E. P. 1593 of 1888.   ‡ Epstein, D. R.-P. 103660 of 1899.
§ Hantzsch and Borghaus, *Ber.*, 1897, **30**, 93.
‖ D. R.-P. 103685 of 1899.
¶ E. P. 17546 of 1892.   ** E. P. 18624 of 1900, D. R.-P., 168299.

In order to diazotize only one amino-group in sulphonated m-diamines, the solution of the base is mixed with the calculated quantity of alkali nitrite and then mineral acid added; by this means the diamine is always in contact with the requisite quantity of free nitrous acid, and the diazotization proceeds smoothly.*

Griess also studied † the action of nitrous acid on p-phenylenediamine, and stated that the principal product of the reaction when carried out in the usual way consisted of aminodiazobenzene chloride, one only of the amino-groups having been diazotized. It was found, however, that by this method a mixture of the diazo- and the tetrazo-compounds was obtained.‡

Later, Griess § was successful in preparing the tetrazo-compound by using the same method as he had employed in the preparation of the m-tetrazobenzene chloride, and the dry tetrazo-sulphate has been obtained in small amount.||

The use of a diazo- and tetrazo-compound derived from p-phenylenediamine has become of very great practical importance in the manufacture of azo-dyes, but as it is essential that a single compound and not a mixture of diazo- and tetrazo- should be prepared, and, further, that no large excess of nitrous acid should be present, these compounds are now prepared indirectly.

For this purpose either p-nitroaniline or p-aminoacetanilide is used as the starting-point. If a compound involving the use of the diazo-chloride is required, the above substances are diazotized in the usual way,¶ and, after coupling with the desired component, the nitro-group is reduced by sodium sulphide solution, or the acetyl group is removed by heating with sodium hydroxide. In each case, if X denotes the component, we obtain the compound

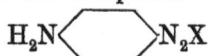

$$\text{H}_2\text{N}\langle\quad\rangle\text{N}_2\text{X}$$

* D. R.-P. 152879.
† Ber., 1884, 17, 697.
‡ Nietzki, Ber., 1884, 17, 1350.
§ Ber., 1886, 19, 317.
|| Hantzsch and Borghaus, loc. cit.
¶ It is singular that, although p-aminoacetanilide, $\text{NH}_2 \cdot \text{C}_6\text{H}_4 \cdot \text{NHAc}$, is very easily diazotized, methyl-p-phenylenediamine, $\text{NH}_2 \cdot \text{C}_6\text{H}_4 \cdot \text{NHMe}$, cannot be thus transformed, nitrogen being evolved even below 0° by the action of nitrous acid. (Hantzsch, Ber., 1902, 35, 896.)

If the tetrazo-compound had been desired, this product is now diazotized in the usual way, and the diazo-compound coupled with a molecule of the same component X, or a different one Y, giving us a dyestuff derived from the tetrazo-compound of p-phenylenediamine, of formula

$$YN_2 \left\langle \right\rangle N_2X$$

In the event of the component X containing an amino-group, and at the same time (as is usual) belonging to the naphthalene series, care is taken to use the calculated quantity of sodium nitrite (one molecule), when the NH_2 group united to the benzene ring is completely diazotized, leaving the other NH_2 group intact. This can generally be also diazotized by using a second molecule of nitrite.

We have seen already (p. 15) that certain substituted amines present difficulties to the diazotizing process, some, in fact, being incapable of diazotization by the usual method. Similar examples occur amongst the substituted diamines: thus o-nitro-p-phenylenediamine cannot be directly converted into the tetrazo-compound, but only the diazo-compound is formed.   Even an excess of nitrite fails to convert more than one amino-group into the diazo-group,* the constitution of the product being in all probability

$$NH_2 \left\langle \overset{\displaystyle NO_2}{} \right\rangle N_2Cl$$

The nitro-p-phenylenediamine is best diazotized by dissolving the hydrochloride in water, adding an excess of acetic acid, and then excess of sodium nitrite at 5–10°. It is very remarkable that if the diazo-compound is coupled with a component such as R salt and an azo-dye formed, the remaining amino-group may now be easily converted into the diazo-group.†   Other instances of this are known in the naphthalene series (see p. 24). Differences in the behaviour of two amino-groups in the substituted benzene molecule had indeed been detected by Griess, who found that p-diamino-

* Bülow, *Ber.*, 1896, **29**, 2285.   † E. P. 6630 of 1892.

benzoic acid yielded p-aminodiazobenzoic acid and not the tetrazo-derivative.*

The diamines of the diphenyl series have attained very great importance owing to their use in the production of dyestuffs which dye cotton without the aid of a mordant. The simplest of these is benzidine,

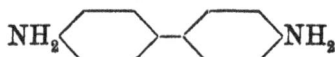

$$NH_2\!\!\left\langle\!\!\bigcirc\!\!\right\rangle\!-\!\left\langle\!\!\bigcirc\!\!\right\rangle\!NH_2$$

which presents no difficulty in undergoing diazotization (contrary to the statement of Kaufler †), the most suitable temperature being 8–10°, and both amino-groups being easily diazotized.

It is also possible to obtain the monodiazo-compound by mixing solutions of benzidine hydrochloride and tetrazodiphenyl chloride and allowing the mixture to remain for two or three days at 10–20°.‡ After filtering off the dark-coloured insoluble by-products, the solution contains principally aminodiazodiphenyl chloride—

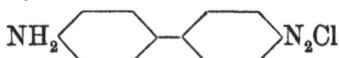

$$NH_2\!\!\left\langle\!\!\bigcirc\!\!\right\rangle\!-\!\left\langle\!\!\bigcirc\!\!\right\rangle\!N_2Cl$$

Other diamines, such as tolidine, dianisidine, ethoxybenzidine, dichlorobenzidine, nitro- and dinitro-benzidine, diaminostilbenedisulphonic acid, are tetrazotized in exactly the same manner as benzidine.

Turning now to the naphthalene series the phenomena observed in the diamines of the benzene series are again present.

Those diamines containing the amino-groups in the ortho or peri positions yield with nitrous acid azimino-compounds (2:3-naphthylenediamine) §. Their sulphonic acids behave similarly.‖

The meta-diamines behave like the m-diamines of the benzene series, giving brown colouring matters.

1:4-Naphthylenediamine is diazotized with still greater difficulty than p-phenylenediamine. Nitrous acid acts in this case also as an oxidizing agent, and 1:4-naphthaquinone is

* *Ber.*, 1884, **17**, 603.
† *Annalen*, 1907, **351**, 151.
‡ Täuber, *Ber.*, 1894, **27**, 2627.
§ *Ber.*, 1894, **27**, 765.
‖ E. P. 8645 of 1895.

formed. In order to obtain the diazo- or tetrazo-derivative the same method is adopted as in the case of p-phenylene-diamine, namely, to convert one amino-group into the acetyl-amino-group and then to diazotize the remaining amino-group.* If this is then coupled with a suitable component, forming an azo-dyestuff, the acetyl group may be split off and the free amino-group now diazotized.

The same procedure is used in preparing the diazo- or tetrazo-derivatives of the 1:4-naphthylenediaminesulphonic acids, except in the case of the acid containing the SO_3H group in the position 2—

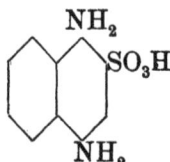

This acid exhibits a great tendency towards the formation of oxidation products when treated with sodium nitrite in the presence of mineral acids, but the diazotization proceeds smoothly when acetic or oxalic acids are used.† It is re-markable that only one amino-group is attacked, and it has been found impossible to prepare a tetrazo-derivative. This behaviour is to be attributed, perhaps, to the protective action of the sulphonic acid group, and consequently the diazo-compound probably possesses the constitution—

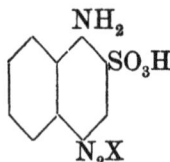

X being the organic acid radical.

If now this diazo-compound is coupled with a phenol or naphthol, the resulting azo-dyestuff is easily diazotized. This behaviour is analogous to that exhibited by o-nitro-p-pheny-lenediamine (p. 22).

It has been found also that if the monoacetyl derivative of

* E. P. 18783 of 1891.
† E. P. 2946 of 1896.

this sulphonic acid is prepared, the formula of which is probably

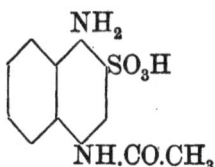

$$\text{NH}_2$$
$$\text{SO}_3\text{H}$$
$$\text{NH.CO.CH}_3$$

the free amino-group readily undergoes diazotization.*

The remaining naphthylenediamines and their sulphonic acids are easily converted into the tetrazo-compounds.†

The transference of triamines into the corresponding diazo-compounds cannot be illustrated by many examples, as cases of this are rare. The best known are probably those of rosaniline and para-rosaniline. These bases, containing, of course, three free amino-groups, were diazotized by Caro and Wanklyn‡ and E. and O. Fischer,§ who thus prepared compounds containing three diazo-groups.

The formation of diazoamino-compounds in this reaction has also been observed. ||

§ **8. 'Solid diazo-compounds'.**—Mention has already been made of the great technical importance of diazotized p-nitro-aniline, owing to its use in the production of the 'para-red' by combination with β-naphthol on the cotton fibre.

In order to enable the dyer to avoid the preparation of this and other diazo-compounds in the dyehouse, several processes have been adopted for the purpose of supplying the users with the diazo-compound ready made.

Such preparations mostly consist of a paste of the diazo-compound in a very concentrated form, or of a sparingly soluble diazo-salt. A remarkable compound, produced by the action of alkalis on the diazo-chloride, which is very stable

---

* E. P. 17064 of 1896.
† E. P. 26020 of 1896; see also Lange, *Chem. Zeit.*, 1888, **12**, 856. Kaufler and Karres (*Ber.*, 1907, **40**, 3263) could only diazotize one amino-group of 2 : 7-naphthylenediamine, using amyl nitrite in alcoholic solution, but patents have been taken out formerly for diazo-dyestuffs from the tetrazo-compound.
‡ *Zeitsch. f. Chem.*, 1866, 511.   § *Annalen*, 1878, **194**, 269.
|| Pelet and Redard, *Bull. Soc. chim.*, 1904 [iii], **31**, 644.

and which yields the diazo-chloride on acidifying, has also been put on the market. The nature of this substance is fully discussed on p. 96.

It is prepared by treating the p-nitrodiazobenzene chloride or other diazo-salts containing nitro- or halogen-groups with caustic alkali at 60–70°.* The diazo-salts prepared from aniline and its homologues are treated at 120°.†

The substances formed may be dried or used as a paste; by the action of a mineral acid the free diazo-chloride is regenerated.

The diazo-compound of p-nitroaniline, after having been treated in this manner with alkali, is known as 'Nitrosamine red in paste'. Another way in which to obtain the diazo-salt in a more stable condition is to mix it with a solution of sodium α-naphthalenesulphonate, ‡ sodium nitrobenzenesulphonate,§ or sodium naphthalenedisulphonates, ‖ and the tetrazo-salts of benzidine, &c., can be condensed with sodium 2-naphthol-3:6:8-trisulphonate, or sodium 2-naphthol-1-sulphonate, when additive compounds, and not azo-dyestuffs, are obtained.¶ Further, the zinc chloride double salts of diazotized aminoazo-compounds are also prepared.** All these stable compounds may be dried.

A simpler method is to diazotize the nitroaniline in a very concentrated solution by passing nitrous acid gas through a solution of p-nitroaniline in sulphuric acid, or even to evaporate the diazo-solution (prepared from sulphuric acid) in a vacuum at a temperature not exceeding 45°. Anhydrous sodium sulphate is now added, which, with the excess of sulphuric acid, is converted into the bisulphate, and the paste, which soon solidifies, may be powdered.†† The substance obtained from diazotized p-nitroaniline in this way is called 'Azophor red P.N.', 'Nitrazol C', Azogen red, and Benzonitrol; and that from diazotized dianisidine 'Azophor blue D'. (For instances of the elimination of groups during diazotization see p. 63.)

---

* E. P. 20605 of 1893.   † E. P. 3397 of 1894, 13460 of 1895.
‡ E. P. 18429 of 1894.   § D. R-P. 88949 of 1894.
‖ D. R-P. 94280 of 1894.   ¶ E. P. 8989 of 1895, 11757 of 1895.
** E. P. 1645 of 1896; D. R-P. 89437 of 1896.
†† E. P. 21227 of 1894; D. R-P. 85387 of 1894; E. P. 15353 of 1897.

CHAPTER III

THE MECHANISM OF THE DIAZOTIZING PROCESS

§ 1. **Thermochemistry.**—The formation of diazo-compounds proceeds with absorption of heat; the reaction is thus an endothermic one.

The development of heat which is observed in the usual process of preparing these compounds is due to the formation of water and sodium chloride.

The values which have been recorded for the heat of formation of diazo-compounds are as follows:—

Diazobenzene nitrate	. . .	-47.4 calories.*
Diazobenzene chloride	. . .	-44.0 „
o-Diazotoluene chloride	. . .	-41.8 „
p-Diazotoluene chloride	. . .	-42.3 „ ‡

§ 2. **Explosibility of dry diazo-compounds.**—From the fact that the formation of diazo-compounds is accompanied by absorption of heat, it was to be expected that these substances would be unstable, and it is found that nearly all diazo-salts are very liable to explode when in the dry state; the most unstable in this respect being those containing nitro-groups. Thus diazobenzene nitrate‡ is more explosive than the sulphate, and a case is on record where p-nitrodiazobenzene nitrate exploded violently when lightly touched with a platinum spatula. §

Great care must be taken, therefore, in handling these substances, as they are extremely unreliable, and may never be regarded as safe. Diazobenzene chloride, usually looked on as comparatively stable, exploded on one occasion, apparently

---

* Berthelot and Vielle, *Compt. rend.*, 1881, **92**, 1076.
† Vignon, *Bull. Soc. chim.*, 1888 [ii], **49**, 906.
‡ Knoevenagel, *Ber.*, 1890, **23**, 2994.
§ Bamberger, *Ber.*, 1895, **28**, 538.

spontaneously, with very great violence;* and a violent explosion of dry diazobenzenesulphonic acid, which had been prepared some years previously, occurred in 1901.† An exactly similar accident befell the author of this book in 1896.

In spite of the danger of working with such substances, determinations of the temperature at which diazo-compounds explode have been made. Thus dry *m*- and *p*-nitrodiazobenzene chlorides explode at 118° and 85° respectively,‡ and diazobenzene nitrate explodes above 90°. §

§ 3. Velocity of diazotization.—The rate at which amines are diazotized has been determined by Hantzsch and Schumann.‖ Diazotization, of course, proceeds with extreme rapidity under ordinary conditions, and the experiments were therefore conducted with *N*/1000 solutions. Using a colorimetric method for estimating the nitrous acid, it was found that, in presence of an excess of acid, the rate of diazotization of aniline, *p*-toluidine, *m*-xylidine, *p*-bromoaniline, and *p*-nitroaniline is the same in each case. Further, if the temperature is raised, the rate is increased.

The reaction which takes place is of the second order, and the values obtained for the velocity-constant—

$$C = \frac{x}{t(a-x)}$$

were 0·036 for aniline, 0·038 for *p*-toluidine, 0·041 for *m*-xylidine, and 0·045 for *p*-bromoaniline in *N*/1000 solution with one molecule of free acid at 0°.

Schumann then measured the velocity by observing the fall of electrical conductivity which takes place during diazotization. ¶ He was able to confirm the previous experiments, and concluded that all aromatic amines are diazotized at approximately the same speed.

* Hantzsch, *Ber.*, 1897, **30**, 2342, footnote.
† Wichelhaus, *Ber.*, 1901, **34**, 11.
‡ Oddo, *Gazzetta*, 1895, **25**, i. 327.
§ Berthelot and Vielle, *Compt. rend.*, 1881, **92**, 1074.
‖ *Ber.*, 1899, **32**, 1691.
¶ *Ber.*, 1900, **33**, 527.

CHAPTER IV

THE REACTIONS OF THE DIAZO-COMPOUNDS

§ 1. **Action of water.**—When a diazo-salt is heated with water a phenol is formed * according to the equation

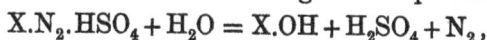

$$X.N_2.HSO_4 + H_2O = X.OH + H_2SO_4 + N_2,$$

X denoting the aromatic nucleus.

The reaction is best carried out in the presence of sulphuric acid. If the diazo-nitrate is used the nitric acid liberated attacks the phenol, forming nitrophenols.

The ordinary method of carrying out the operation is to diazotize in presence of sulphuric acid and then to add, if necessary, a further quantity of sulphuric acid. The solution is then either directly boiled until no further evolution of nitrogen takes place, or steam may be passed into the solution, or the solution may be added slowly to boiling dilute sulphuric acid.

In one or other of these ways most diazo-compounds yield the corresponding phenol, which is isolated by the usual means. For example, 4:4'-dihydroxydiphenyl is obtained in the following manner. 25 grams of benzidine are dissolved by the aid of heat in 500 c.c. of water and 30 c.c. of concentrated hydrochloric acid. The solution is cooled to 5° by adding ice, and then 18 grams of sodium nitrite dissolved in a small quantity of water are poured in slowly, the temperature not being allowed to rise above 10°; this is effected by adding more ice if necessary.

100 grams of concentrated sulphuric acid are now added, and steam passed into the mixture until crystals of dihydroxydiphenyl begin to separate out and the solution gives no further colour with an alkaline solution of R salt or β-naphthol. On cooling, the precipitate is filtered, dissolved in dilute caustic soda, the solution filtered from any insoluble matter, and

* Griess, *Annalen*, 1866, **137**, 67.

reprecipitated by hydrochloric acid. The dihydroxydiphenyl is recrystallized from dilute alcohol when it is obtained pure.*

The process is carried out on the large scale in the manufacture of several naphthol- and dihydroxynaphthalene-sulphonic acids, and a classical example of this decomposition, as applied to diazo-compounds derived from triamines, is the production of aurin from para-rosaniline. †

Owing to the very great reactivity of the diazo-salts and their well-known capacity of coupling or combining with phenols, it is obvious that there is a great tendency for secondary reactions to take place, interfering, to a certain extent, with the quantitative production of the hydroxy-compounds; further, owing to the extreme differences in the relative stability of diazo-salts, other more obscure side reactions are liable to intervene, particularly in those cases where the decomposition can be completed only by long heating.

Secondary reactions occur indeed even in the simplest case ; thus in the decomposition of diazobenzene sulphate a small quantity of hydroxydiphenyl is formed in consequence of the action of some of the undecomposed diazo-compound on phenol. ‡

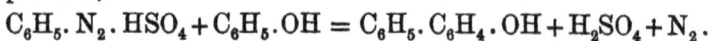

$$C_6H_5 . N_2 . HSO_4 + C_6H_5 . OH = C_6H_5 . C_6H_4 . OH + H_2SO_4 + N_2 .$$

In very many cases also, particularly in the naphthalene series, the solution becomes deeply coloured owing to the coupling of the diazo-compound with the naphthol formed, with production of the azo-dyestuff.§ Thus in the decomposition of diazo-α-naphthalene-4-sulphonic acid this reaction invariably occurs, even in the presence of sulphuric acid; consequently a large excess of acid is usually taken in order to limit this formation of colouring matter as far as possible.

There are many cases recorded in the literature where it has been found impossible to obtain even a trace of an hydroxy-compound by carrying out the decomposition in the manner described above. Most of these occur among extremely stable diazo-compounds, such as those derived from the halogen or nitro-substituted amines.‖

* Compare also Hirsch, *Ber.*, 1889, **22**, 335.
† *Annalen*, 1878, **194**, 301.   ‡ Hirsch, *Ber.*, 1890, **23**, 3705.
§ *Trans.*, 1903, **83**, 221.   ‖ *Amer. Chem. J.*, 1889, **11**, 319.

Recognizing that the cause of this might be due to an insufficiently high temperature, Heinichen * adopted the method of heating the strong diazo-solution with concentrated sulphuric acid, whereby the boiling-point becomes raised to 150°. In this way he obtained 2 : 6-dibromophenol from the corresponding diazo-salt after the usual method had failed.

As, however, the stability of diazo-compounds has been shown to increase with addition of sulphuric acid,† this method is not always successful.

A novel way of attacking the problem is that described by Kalle & Co. ‡ The non-production of phenols in certain cases being evidently due, as already indicated, to condensation between the diazo-compound and the phenol formed, any process depending on the removal of the latter when set free would be expected to stand more chance of success. The method adopted by this firm is therefore to carry out the decomposition by dropping the diazo-solution into a mixture of dilute sulphuric acid and sodium sulphate heated to 135–145°, and allowing any volatile products to distil over.§ In this way the temperature is kept high without using concentrated sulphuric acid.

By this means a good yield of guaiacol is obtained from the diazo-salt of o-anisidine,‖ and the diazo-salts of s-tribromo- and s-trichloro-aniline, which under no other conditions could be made to yield phenols, gave a small yield of s-tribromophenol and s-trichlorophenol respectively.¶ Another method of procedure is to decompose the diazo-compound at the moment of its formation by adding a solution of sodium nitrite to a boiling solution of the base in hydrochloric acid. In this way a good yield of p-nitro-o-cresol can be obtained from p-nitro-o-toluidine, but if the decomposition is carried out in the usual way internal condensation occurs, with

* *Annalen*, 1889, **253**, 281.   † *Ber.*, 1905, **38**, 2511.
‡ E. P. 7233 of 1897.
§ An alternative method consists in adding the diazo-solution to a boiling 50 per cent. aqueous solution of copper sulphate. (D. R.-P. 167211, *Soc. Chim. des Usines du Rhône.*)
‖ The diazo-sulphate of p-anisidine yields quinol on heating with water to 140°. (Salkowski, *Ber.*, 1874, **7**, 1008.)
¶ Cain, *Trans.*, 1906, **87**, 19.

formation of nitroindazole.* Certain derivatives of o-anisidine, the diazo-compounds of which have failed to yield the
corresponding substituted guaiacol, would, in all probability,
behave in the normal manner if one of the above methods
were applied.†

Various other apparent exceptions to the general rule have
been described by Wroblewski,‡ who obtained the substituted
hydrocarbons only, and not the phenols, from the diazo-salts
derived from dibromoaniline, dibromo-p-toluidine, and bromo-
and chloro-p-toluidine. These diazo-salts have been recently
examined, with the result that, in each case, the corresponding
phenol was obtained.§ Wroblewski's results were probably
due, as indeed he himself suggests,|| to the presence of alcohol
used in the preparation. The production of dibromophenetole
by heating the diazo-compound of dibromo-o-phenetidine ¶ is
probably explained in the same way.

A case which does not seem to accord with this explanation
is that of ethyl diazogallate, which can be crystallized from
water, and when heated with water in a sealed tube for four
hours to 220° yields ethyl gallate, the nitrogen having been
completely eliminated.**

By treating aminoindazole with nitrous acid and warming
the resulting diazo-compound with water, Bamberger †† prepared a new class of diazo-compounds, to which the name
'triazolens' is given. He formulated the compound according
to the equation

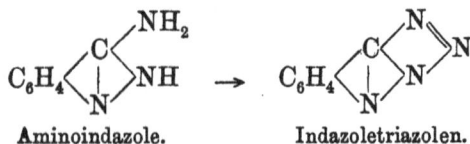

$$C_6H_4 \left\langle \begin{array}{c} \overset{NH_2}{C} \\ \underset{N}{|} NH \end{array} \right. \quad \rightarrow \quad C_6H_4 \left\langle \begin{array}{c} \overset{N}{C} \\ \underset{N}{|} N \end{array} \right.$$

Aminoindazole.   Indazoletriazolen.

but Hantzsch ‡‡ regarded it as a diazide—

* Witt, Noelting, and Grandmougin, Ber., 1890, **23**, 3635.
† Meldola, Woolcott, and Wray (Trans., 1896, **69**, 1327) obtained
resins only by boiling the diazo-salts of p-bromo- and p-nitro-o-anisidine
with water or dilute sulphuric acid.
‡ Ber., 1874, **7**, 1061.
§ Cain, Trans., 1906, **89**, 19. || Ber., 1884, **17**, 2704.
¶ Möhlau and Oehmichen, J. pr. Chem., 1881 [ii], **24**, 476.
** Power and Sheddon, Trans., 1902, **81**, 77.
†† Ber., 1899, **32**, 1773.   ‡‡ Ber., 1902, **35**, 89.

$$C_6H_4 \diagdown \begin{matrix} -C \diagup N \\ \diagdown N \\ N:N \end{matrix}$$

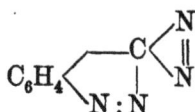

In the diphenyl series certain exceptions to the general rule have been observed. On heating the tetrazo-salts prepared from dianisidine and 3 : 3'-dichlorobenzidine

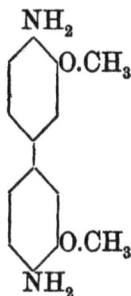

with dilute sulphuric acid, no phenol whatever was obtained,[*] the products being apparently of a quinonoid character. By using Heinichen's method a small amount of the dihydroxy-compound was obtained only in the latter case. An attempt to replace both the diazo-groups in ethoxytetrazodiphenyl sulphate

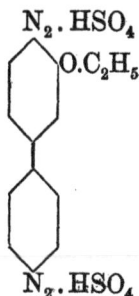

led to an interesting result.[†] It was found that the diazo-groups varied greatly in stability, and an intermediate product was isolated having the formula

---

[*] Cain, *Trans.*, 1903, **83**, 688.
[†] Cain, *Trans.*, 1905, **87**, 5.

$$N_2 . HSO_4$$

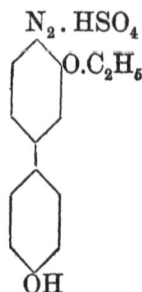

O.C$_2$H$_5$

OH

§ 2. **Stability of diazo-solutions.**—Very great differences
occur between the various diazo-salts with regard to their
power of resisting decomposition by water. Many decompose
rapidly at the ordinary temperature, whilst others remain
apparently unchanged after prolonged boiling.

Several cases of great stability are described by Griess;
most of them occur among the halogen or nitro-substituted
diazo-salts. Experiments of a somewhat qualitative character
were performed by Oddo,* who diazotized a number of amines
at various temperatures, and determined the quantity of the
diazo-compound formed. He found that at 100–105° much
diazo-compound is produced from *m*- and *p*-nitroaniline,
p-chloroaniline, 1:3:4-dinitroaniline, and 1:2:5- and 1:3:6-
nitrotoluidine, whilst little is obtained with *m*-chloro- and
bromo-aniline, *o*-nitroaniline, *p*-aminobenzoic acid, and 1:5:2-
nitrotoluidine. At 80–85° much diazo-compound is obtained
with the substances named above as giving little at 100–105°,
whilst small yields are furnished by aniline, *p*-toluidine, and
a- and *β*-naphthylamine; at 60–65° the four bases last named
give good yields of diazo-compounds, whilst *o*-toluidine and
p-xylidine give very poor ones. At 40–45° *p*-xylidine in
turn gives a good yield of diazo-compound.

A considerably more exact method of determining the
stability consists in titrating from time to time a portion
of a diazo-solution with a fixed amount of sodium *β*-naph-
tholsulphonate (Schäffer's salt) solution.† The increasing
amount of diazo-solution required to combine with the whole

* *Gazzetta*, 1895, **25**, i. 327; 1896, **26**, ii. 541.
† Hirsch, *Ber.*, 1891, **24**, 324.

of the naphthol solution is a measure of the advance of the decomposition.

The main results obtained in this way are given in the table on p. 37.

This titration method has been used by some later investigators,* but it is not suitable for exact measurements, owing to the possibility of secondary reactions taking place between the diazo-compound and the alkali or sodium acetate, which must be added to the naphtholsulphonic acid in order to effect complete combination. (For an account of this secondary reaction see p. 96.)

Hausser and Muller † introduced an entirely different method from the foregoing. They heated solutions of various diazo-compounds at fixed temperatures and measured the volume of nitrogen evolved.

The decomposition belongs to the class of unimolecular processes, and is represented by the well-known expression—

$$C = \frac{1}{t} \log \frac{A}{A - x}$$

By comparing the values of the constant obtained the relative stability of the diazo-compounds may be determined.

A constant value was obtained only in the case of the diazo-salts from sulphanilic acid and p-toluidinesulphonic acid. In the case of the other amines examined the values for C were not constant, and from these results somewhat erroneous conclusions were drawn. Hantzsch, ‡ using the same method, measured the rate of decomposition of the diazo-chlorides prepared from aniline, p-bromoaniline, p-toluidine, p-anisidine, and ψ-cumidine, and showed that at 25° all these substances gave a constant value for

$$\frac{1}{t} \log \frac{A}{A - x}.$$

It must be noted that the experiments of Hausser and Muller and of Hantzsch were carried out with solutions pre-

* Buntrock, *Leipziger Monatsschrift für Textil-Industrie*, 1898, 608; Schwalbe, *Zeitsch. Farb.-Ind.*, 1905, 4, 433.

† *Bull. Soc. chim.*, 1892 [iii], 7, 721; 1893, 9, 353. *Compt. rend.*, 1892, 114, 549, 669, 760, 1438.

‡ *Ber.*, 1900, 33, 2517.

pared by dissolving the dry diazo-salt in water. Recent investigations have shown * that solutions of diazo-salts prepared directly from the following amines—aniline, the toluidines, sulphanilic acid, the nitroanilines, p-aminoacetanilide, 3:3'-dichlorobenzidine, α- and β-naphthylamine, and a number of sulphonic acids derived from the two last, decompose in accordance with the above formula. This holds good at temperatures ranging from 20° to 100°. The diazo-salt prepared from m-toluidine is the most unstable of those examined, the value of C at 20° being 0·00208. o-Toluidine comes next with a value of C of 0·00187; aniline gives a value for C of 0·00072.

The diazo-salts of the nitroanilines are extremely stable, the ortho-compound being the most and the para- the least stable. The value of C for o-nitrodiazobenzene chloride is 0·00555 at 100°. Exceptions to the rule are shown by certain tetrazo-salts and those diazo-salts which are insoluble in water.

The rate of decomposition of diazo-salts increases rapidly with the temperature,† the values of C obtained being in accordance with Arrhenius's formula for the temperature coefficient, namely

$$Ct_1 = Ct_0 e^{A(T_1 - T_0) : T_1 T_0}.$$

The rate of decomposition (in the case of diazobenzene salts) is independent of the quantity of mineral acid present (except sulphuric acid, which tends to withdraw water from the sphere of action), and is independent of the nature of the acid. Equivalent solutions of diazobenzene chloride, bromide, sulphate, nitrate, and oxalate decompose at the same rate. ‡

The presence of colloidal platinum or silver increases the rate of the decomposition, owing to catalytic action. § Finally, it may be useful to append a table showing the relative

---

* Cain and Nicoll, *Trans.*, 1902, **81**, 1412; 1903, **83**, 206.

† Cain and Nicoll, *Trans.*, 1903, **83**, 470; Euler, *Annalen*, 1902, **325**, 292.

‡ Cain, *Ber.*, 1905, **38**, 2511; Euler, loc. cit.

§ Euler, *Öfversigt af Kongl. Vetenskaps. Akad. Förhandl. Stockholm*, 1902, No. 2, 227. Compare also Schwalbe, *Ber.*, 1905, **38**, 2196, 3071; Cain, *Ber.*, 1905, **38**, 2511.

stability of various diazo-salts as determined by various observers. The amine giving the most stable diazo-salt is at the top.

Cain and Nicoll.	Euler.	Hantzsch.	Hirsch.	Oddo.
o-Nitroaniline	o-Chlorobenzene	p-Anisidine	Sulphanilic acid	m-Nitroaniline
m-Nitroaniline	o-Anisidine	p-Bromoaniline	Aniline	p-Nitroaniline
p-Nitroaniline	p-Anisidine	p-Toluidine	m-Xylidine	p-Chloroaniline
Sulphanilic acid	p-Chloroaniline	Aniline	o-Toluidine	1:3:4-Dinitroaniline
p-Toluidine	p-Bromoaniline	ψ-Cumidine		1:2:5-Nitrotoluidine
Aniline	m-Chloroaniline			1:3:6-Nitrotoluidine
o-Toluidine	p-Toluidine			m-Chloroaniline
m-Toluidine	m-Bromoaniline			m-Bromoaniline
	p-Aminobenzoic acid			o-Nitroaniline
	m-Aminobenzoic acid			p-Aminobenzoic acid
	Aniline			1:5:2-Nitrotoluidine
	ψ-Cumidine			Aniline
	o-Toluidine			p-Toluidine
	m-Toluidine			α-Napthylamine
				β-Napthylamine
				o-Toluidine
				p-Xylidine

CHAPTER V

THE REACTIONS OF THE DIAZO-COMPOUNDS
(continued)

§ 1. Action of alcohols.—The action of alcohol on diazo-compounds was, of course, studied by Griess, who obtained benzene from diazobenzene salts, and dinitrophenol from diazodinitrophenol.

The production of the hydrocarbon or complete elimination of the diazo-group by the action of boiling alcohol was, for many years, regarded as a general reaction, in spite of the observation of Wroblewski,[*] who found that the diazo-salt of chlorotoluidine gave, not the chloro-hydrocarbon, but the corresponding chlorophenetole. Four years later also Hayduck[†] showed that when o-toluidinesulphonic acid was diazotized, and the resulting diazo-salt boiled with alcohol, phenetolesulphonic acid was obtained.

A striking application of the reaction was made by E. and O. Fischer in 1878,[‡] who showed that when the diazo-compound of paraleucaniline was boiled with alcohol the three diazo-groups were eliminated, with formation of triphenyl-methane ; diazo-leucaniline, in the same way, gave tolyl-diphenylmethane.

When the bisdiazo-derivative of benzidine is warmed with ethyl alcohol to 40–45° only one diazo-group is eliminated, the second requiring a higher temperature for its removal—

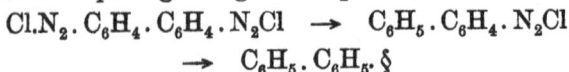

$$Cl.N_2 . C_6H_4 . C_6H_4 . N_2Cl \rightarrow C_6H_5 . C_6H_4 . N_2Cl$$
$$\rightarrow C_6H_5 . C_6H_5.\ [\S]$$

Examples of the formation of ethers in this reaction were, however, rapidly multiplying, amongst which may be mentioned the cases of m-aminobenzenedisulphonic acid,[||] cumi-

---

* *Ber.*, 1870, **3**, 98.　　　　† *Annalen*, 1874, **172**, 215.
‡ *Annalen*, 1878, **194**, 242.　　　§ *Ber.*, 1898, **31**, 479.
|| Zander, *Annalen*, 1879, **198**, 1.

dinesulphonic acid,* aminotetramethylbenzene, cumidine,†
and o-toluidinedisulphonic acid,‡ all of which yielded, when
diazotized and then treated with alcohol, the corresponding
ethyl ethers.

The reaction may thus proceed in two ways, according to
the following equations :—

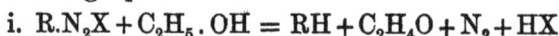

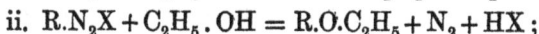

i. $R.N_2X + C_2H_5 . OH = RH + C_2H_4O + N_2 + HX$

ii. $R.N_2X + C_2H_5 . OH = R.O.C_2H_5 + N_2 + HX$;

where R denotes a hydrocarbon radical and X an acid
radical. A systematic investigation into the whole question
was next undertaken by Remsen § and his pupils, and it was
very quickly demonstrated that the normal reaction is the
formation of ethers in accordance with the second of the
foregoing equations. ‖ The course of the reaction is, however, somewhat complicated, and depends on many factors,
such as the position and nature of the substituents, the
pressure at which the operation is carried on, &c.

§ **2. Influence of substituents.**—The presence of the acid
radicals, CO_2H, Cl, Br, NO_2, &c., tends to induce the complete
elimination of the diazo-group, and this influence is greatest
when these radicals are in the ortho-position with respect to
the diazo-group; their influence is less in the meta-position
and least in the para-position.

Thus of the chlorodiazobenzene nitrates, the ortho- and
meta-compounds yield only chlorobenzene with ethyl alcohol,
but the para-compound gives rise to the formation of a little
p-chlorophenetole.¶ In the case of the diazobenzoic acids the
ortho-compound gives benzoic acid only, whilst the meta- and
para- yield the alkyloxy-derivatives. **

Another interesting example is that of the nitrodiazo-
benzene salts. When heated with methyl alcohol the ortho-

* Haller, *Ber.*, 1884, **17**, 1887.
† Hofmann, *Ber.*, 1884, **17**, 1917.
‡ Hasse, *Annalen*, 1885, **230**, 286.
§ *Ber.*, 1885, **18**, 65.
‖ Remsen and Palmer, *Amer. Chem. J.*, 1886, **8**, 243.
¶ Cameron, *Amer. Chem. J.*, 1898, **20**, 229.
** Remsen and Orndorff, *Amer. Chem. J.*, 1887, **9**, 387. Compare also
Griess, *Ber.*, 1888, **21**, 978.

compound gives 87 per cent. of the theory of nitrobenzene;
from the meta-compound 51 per cent. is obtained, together
with a little *m*-nitroanisole, whilst the para-derivative gives
about 40 per cent. of nitrobenzene and 8 to 17 per cent. of
p-nitroanisole.

In the naphthaline series the 1 : 2-, 2:1-, and 1 : 4-nitrodiazo-
naphthalene sulphates all yield nitronaphthalene with ethyl
alcohol,* whereas the ethoxy-derivative is obtained from both
a- and *β*-diazonaphthalene sulphates.†

§ 3. **Influence of the alcohol used.**—The tendency towards
the formation of hydrocarbons is increased as the molecular
weight of the alcohol increases. Diazobenzene chloride and
sulphate with methyl alcohol yield anisole as the sole product,
no benzene being formed.‡   With ethyl alcohol the chief product is phenetole, but a little benzene is also obtained. *o*-Diazo-
toluene sulphate with methyl alcohol yields tolyl methyl ether
and only a trace of toluene.§

The diazo-salts of *m*-chloro- and *m*-bromo-aniline also yield
only the corresponding halogen derivatives of benzene when
heated with ethyl alcohol, but when methyl alcohol is used
the chief product in each case is the halogenated anisole, only
small quantities of chloro- and bromo-benzene being produced.
The diazo-sulphates of *p*-chloro- and *p*-bromo-aniline illustrate
this point very clearly; with ethyl alcohol no ethers are obtained, whilst with methyl alcohol the ethers are the sole
products.‖   The higher alcohols behave in a similar manner
to methyl and ethyl alcohols. With diazobenzene chloride
n- and *iso*-propyl alcohol yield phenyl propyl ethers but no
trace of propaldehyde or acetone; amyl alcohol gives both
phenyl amyl ether and valeraldehyde or its condensation products; and benzyl alcohol gives benzaldehyde with only a
little phenyl benzyl ether. Glycerol behaves like propyl
alcohol, giving the monophenyl ether, whilst mannitol and
benzoin are not attacked.¶

* Orndorff and Cauffman, *Amer. Chem. J.*, 1892, **14**, 45.
† Orndorff and Kortright, *Amer. Chem. J.*, 1891, **13**, 153.
‡ Hantzsch and Jochem, *Ber.*, 1901, **34**, 3337.
§ Bromwell, *Amer. Chem. J.*, 1897, **19**, 561.   ‖ Cameron, loc. cit.
¶ Hantzsch and Vock, *Ber.*, 1903, **36**, 2061. Compare also Orndorff
and Hopkins, *Amer. Chem. J.*, 1893, **15**, 518.

Phenol acts similarly, thus, when a solution of diazobenzene sulphate is warmed with phenol, diphenyl ether is obtained.* In alkaline solution, however, an azo-compound is produced (see p. 86).

§ 4. **Influence of temperature and pressure.**—The influence of these factors in the decomposition is somewhat difficult to separate, as when the pressure is varied, the boiling-point of the solvent changes.

In the case of the diazo-compound prepared from *p*-toluidine-*o*-sulphonic acid,† the decomposition proceeds slowly at the ordinary pressure. When this pressure is raised by 500 m.m. an almost quantitative yield of the ethoxy-compound is obtained, but below this pressure the yield is diminished, as shown by the following numbers :—

Pressure in mm.	800	700	600	500	400	300	210	120
Ethoxy-compound per cent.	69·8	63·2	57·7	52·8	48·7	43·4	40·6	37·2

With methyl alcohol the methoxy-compound is obtained, and alteration of pressure has no influence on the course of the reaction.‡

From a large number of cases, however, which have been examined, involving the use of both methyl and ethyl alcohol, it is found that the yield of alkyloxy-derivative increases with the pressure.§

§ 5. **Influence of other substances.**—If the decomposition is carried out with the addition of sodium ethoxide, sodium hydroxide, potassium carbonate, or zinc dust, a remarkable effect is produced. The alkyloxy-formation is almost entirely inhibited, and the reaction proceeds mainly with elimination of the diazo-group. Thus in the case of *p*-diazotoluene nitrate and sulphate the ordinary treatment with methyl alcohol results in the production of a good yield of the methoxy-derivative. When, however, sodium methoxide or any of the

* Hofmeister, *Annalen*, 1871, **159**, 191.
† Remsen and Palmer, *Amer. Chem. J.*, 1886, **8**, 243 ; Remsen and Dashiell, ibid., 1893, **15**, 105.
‡ Parks, *Amer. Chem. J.*, 1893, **15**, 320.
§ Shober, *Amer. Chem. J.*, 1893, **15**, 379 ; Metcalf, ibid., 301 ; Beeson, ibid., 1894, **16**, 235 ; Shober and Kiefer, ibid., 1895, **17**, 454 ; Chamberlain, ibid., 1897, **19**, 531.

above substances are present no alkyloxy-compound is obtained, but the product consists mainly of toluene.*

The rule holds good also for tetrazo-compounds of the diphenyl series; thus the tetrazo-chloride of o-ditolyl gives, with methyl alcohol, dimethoxy-o-ditolyl, and with ethyl alcohol a mixture of diethoxy-m-ditolyl and m-ditolyl; but in the presence of sodium methoxide, hydroxide, or zinc dust, no alkylated compound is formed.†

§ 6. **Other methods of reduction.**—The reduction of diazo-salts to the corresponding hydrocarbon may, of course, be effected by reducing agents instead of alcohol: thus Baeyer and Pfitzinger ‡ introduced the method of reducing the diazo-salt to the hydrazine with stannous chloride, and removing the group NH.NH₂, by oxidation with boiling cupric sulphate solution, and by treating diazobenzene formate with stannous formate solution, benzene, together with a little diphenyl, &c., is produced.§

$$C_6H_5 . N_2Cl + SnCl_2 + H_2O = C_6H_6 + N_2 + SnOCl_2 + HCl.$$

The reduction is also effected by adding sodium stannite to a solution of a diazo-compound in sodium hydroxide, ‖

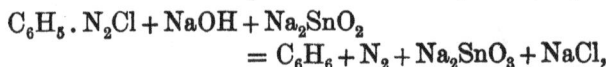

$$C_6H_5 . N_2Cl + NaOH + Na_2SnO_2$$
$$= C_6H_6 + N_2 + Na_2SnO_3 + NaCl,$$

by the use of hypophosphorous acid,¶ an alkaline solution of sodium hyposulphite,** and also when diazides of sulphonic acids are boiled with copper powder and formic acid.††

---

* Beeson, loc. cit.; Chamberlain, loc. cit.; Griffin, *Amer. Chem. J.*, 1897, **19**, 163; Moale, ibid., 1898, **20**, 298.
† Winston, ibid., 1904, **31**, 119.   ‡ *Ber.*, 1885, **18**, 90, 786.
§ Gasiorowski and Waijss, *Ber.*, 1885, **18**, 337; Culmann and Gasiorowski, *J. pr. Chem.*, 1889 [ii], **40**, 97.
‖ Friedländer, *Ber.*, 1889, **22**, 587.   Compare also Eibner, *Ber.*, 1903, **36**, 813.
¶ Mai, *Ber.*, 1902, **35**, 162.   ** Grandmougin, *Ber.*, 1907, **40**, 858.
†† *Ber.*, 1890, **23**, 1632.

CHAPTER VI

THE REACTIONS OF THE DIAZO-COMPOUNDS
(continued)

§ 1. Replacement of the diazo-group by the halogens.

1. **Chlorine.** — Although chloro-derivatives are obtained when a diazo-salt is heated with concentrated hydrochloric acid,* the yield is usually very poor,† and Griess observed that the replacement was more successful when the platinichloride of the diazo-compound was heated with sodium hydroxide, thus—

$$(C_6H_5N_2)_2PtCl_6 = 2C_6H_5Cl + Pt + 2Cl_2 + N_2 .$$

A much more convenient method, however, was introduced by Sandmeyer in 1884.‡

In investigating the action of cuprous acetylide on diazobenzene chloride, he noticed that chlorobenzene was produced, and showed that this was due to the cuprous chloride formed during the reaction.

The replacement is carried out by adding the diazo-solution to a boiling 10 per cent. solution of cuprous chloride in hydrochloric acid. Nitrogen is evolved and the mass distilled with steam when chlorobenzene passes over.

The cuprous chloride may be prepared by heating to boiling a mixture of copper sulphate (250 parts), sodium chloride (120 parts), and water (500 parts). Concentrated hydrochloric acid (1,000 parts) and copper turnings (130 parts) are now added, and the temperature maintained until the mixture loses its colour. The solution is decanted from any undissolved copper and the weight made up to 2,036 parts by the addition of

* Griess, *Ber.*, 1885, **18**, 960.
† When, however, a-diazoanthraquinone is treated with hydrochloric acid, the chloro-derivative is readily obtained, and the presence of cuprous salts is not essential (D. R-P. 131538).
‡ *Ber.*, **17**, 1633, 2650; 1885, **18**, 1492, 1496; 1890, **23**, 1880; see also *Ber.*, 1886, **19**, 810; 1890, **23**, 1628; *Annalen*, 1893, **272**, 141.

concentrated hydrochloric acid. A 10 per cent. solution of cuprous chloride is obtained, which is preserved in an atmosphere of carbon dioxide.*

Gattermann then demonstrated † that the addition of very finely-divided copper to a solution of the diazo-chloride in hydrochloric acid effected the replacement at the ordinary temperature. Gattermann used copper precipitated from copper sulphate solution with zinc, but Ullmann has shown that the 'copper bronze' of commerce may be used equally effectively.‡

A modification of this method consists in using copper sulphate solution to which is added hydrochloric acid and sodium hypophosphite.§

The 'Sandmeyer' reaction, as it is usually called, is considered to be accompanied by the intermediate formation of a compound of the diazo-chloride with the cuprous chloride, and it is important in carrying out this operation that the possibility of the formation of phenols and azo-compounds should be avoided as far as possible. The production of a phenol is due to the decomposition of the diazo-compound before it has been converted into the cuprous chloride compound, or if it is added too slowly to the latter.

According to Erdmann,‖ the normal decomposition of the diazo-cuprous chloride compound takes place rapidly and smoothly only above a certain temperature, which is different for each compound ; these temperatures are about 0°, 27°, and 30–40° in the case of the cuprous chloride derivatives of diazobenzene, o-diazotoluene, and p-diazotoluene respectively. Below these points, the evolution of nitrogen takes place too slowly and is incomplete, part of the diazo-cuprous chloride compound being reduced to an azo-compound by the liberated cuprous chloride. It has been found that the quantity of cuprous chloride required may be reduced to 1/21 and 1/28 molecule per molecule of amine without appreciably reducing the yield of chlorobenzene and m-chloronitrobenzene respectively.¶ This

* Feitler, Zeitsch. physikal. Chem., 1889, 4, 68.
† Ber., 1890, 23, 1218; 1892, 25, 1091.
‡ Ber., 1896, 29, 1878.   § Angeli, Ber., 1891, 24, 952.
‖ Annalen, 1893, 272, 141.
¶ Votoček, Chem. Zeit. Rep., 1896, 20, 70.

is considered to be due to the diazobenzene chloride becoming
first reduced to phenylhydrazine by the cuprous chloride, which
then becomes cupric chloride ; the phenylhydrazine is then
oxidized in presence of hydrochloric acid to chlorobenzene by
the cupric chloride, and the cuprous chloride would then be
re-formed to play the same part again. In confirmation of
this explanation, it is found that phenylhydrazine is oxidized
under the conditions named to chlorobenzene by both cupric
and ferric chlorides, but no phenylhydrazine can be detected
in the Sandmeyer reaction, owing possibly to its momentary
existence. It is also worthy of note that a copper salt is not
necessary in the preparation of iodobenzene by this method,
and this may be due to the fact that hydriodic acid is itself
a reducing agent.*

Still another variation of Sandmeyer's method consists in
electrolysing a solution of a diazo-compound to which cupric
chloride has been added. A thick copper wire is used as
the anode, and a cylinder of sheet copper as the cathode ;
with a current density of 2·1 amperes per sq. dcm., and an
E. M. F. of 10 volts, nitrogen was evolved, and a yield of 64
per cent. of the theory of chlorobenzene was obtained.†

In certain cases the reaction takes a different course from
that already described; thus Gattermann ‡ found that two
benzene nuclei could condense to form diphenyl derivatives,
and the reaction has been extended by Ullmann, who has
prepared a large number of diphenyl compounds by acting on
nitrodiazo-compounds with cuprous chloride.§

2. **Bromine.**—The diazo-group is replaced by bromine in the
same manner as by chlorine.‖ In Griess's method a perbromide
is obtained by adding hydrobromic acid and bromine water to
the diazo-compound,¶ and this on being boiled with alcohol
yields the bromo-derivative thus—

* Walter, *J. pr. Chem.*, 1896 [ii], **53**, 427.
† Votoček and Zenišek, *Zeitsch. Elektrochem.*, 1899, **5**, 485.
‡ *Ber.*, 1890, **23**, 1226.
§ *Ber.*, 1901, **34**, 3802 ; D. R–P. 126961. See also p. 61.
‖ *Phil. Trans.*, 1864, **154**, 673 ; *Annalen*, 1866, **137**, 49.
¶ If a diazophenol is used, a bromodiazophenol is formed ; cp. *J. pr.
Chem.*, 1881 [ii], **24**, 449 ; *Annalen*, 1886, **234**, 1 ; and it is remarkable
that diazosulphanilic acid is indifferent to bromine (Armstrong, *Proc.*,
1899, **15**, 176).

$$C_6H_5 . N_2 . NO_3 + HBr + Br_2 = C_6H_5 . N_2Br.Br_2 + HNO_3$$
$$C_6H_5 . N_2Br.Br_2 + C_2H_5 . OH$$
$$= C_6H_5Br + N_2 + 2HBr + CH_3 . CHO.$$

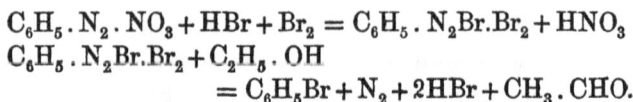

In addition to bromobenzene, p-bromophenetole is formed (cp. p. 38); when ether or glacial acetic acid is used instead of alcohol, bromobenzene alone is produced.* The platinibromide of the diazo-bromide may also be treated in the same way as the platinichloride.

In Sandmeyer's reaction cuprous bromide is substituted for cuprous chloride.

In order to prepare β-bromonaphthalene, Oddo † modified Gattermann's process as follows : 14·3 grams of β-naphthylamine are diazotized and added to a mixture of 36 grams of potassium bromide with 100 grams of water and 30 grams of moist copper powder previously heated to 50-70°. The whole is heated in a reflux apparatus for 15 minutes, and then distilled in steam. A yield of 46-48 per cent. of the theory is obtained.

3. Iodine.—Iodo-derivatives are easily prepared from the diazo-compounds by treating the latter with hydriodic acid. A solution of a little more than the theoretical quantity of sodium or potassium iodide is added to the solution of the diazo-chloride or sulphate. After standing and warming until the evolution of nitrogen has ceased, the liquid is usually made alkaline and the iodo-compound, if it is volatile, distilled with steam. In other cases it may be filtered off.

4. Fluorine.—The diazo-group may be replaced by fluorine by treating the diazo-salt with a solution of hydrogen fluoride in water.‡

The substitution has also been effected from diazoaminocompounds. Thus, if diazoaminobenzene is added to fuming hydrofluoric acid, fluorobenzene is produced,§ and on mixing diazobenzene piperidide (from diazobenzene nitrate and piper-

---

* Saunders, *Amer. Chem. J.*, 1891, **13**, 486.
† *Gazzetta*, 1890, **20**, 631.
‡ *Ber.*, 1879, **12**, 581; 1889, **22**, 1846.
§ Schmitt and Gehren, *J. pr. Chem.*, 1870 [ii], **1**, 395.

idine *) with concentrated hydrofluoric acid, fluorobenzene is formed, thus—

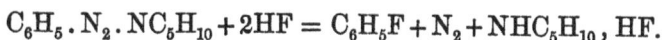

$$C_6H_5 . N_2 . NC_5H_{10} + 2HF = C_6H_5F + N_2 + NHC_5H_{10} , HF.$$

The Sandmeyer reaction is considered by Hantzsch and Blagden † to be a somewhat complicated one, the final result being due to the simultaneous effect of three concurrent actions, namely, (1) the formation of a labile (diazonium, see p. 133) cuprous double salt, which then decomposes in such a way that the radical originally attached to the copper migrates to the aromatic nucleus ; (2) a catalytic action, which is the main action when copper powder is used, whereby nitrogen is eliminated from the diazo-salt, and the acid radical becomes united with the aromatic nucleus ; (3) the formation of azo-compounds, the cuprous being oxidized to a cupric salt.

The first two reactions proceed when p-bromodiazobenzene bromide is subjected to the action of cuprous chloride dissolved in methyl sulphide. The product consists chiefly of chloro-p-bromobenzene mixed with a little p-dibromobenzene.

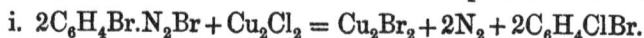

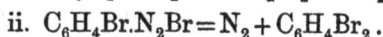

i. $2C_6H_4Br.N_2Br + Cu_2Cl_2 = Cu_2Br_2 + 2N_2 + 2C_6H_4ClBr.$

ii. $C_6H_4Br.N_2Br = N_2 + C_6H_4Br_2 .$

When cuprous bromide is allowed to react with p-bromodiazobenzene chloride, p-dibromobenzene and a little chloro-p-bromobenzene are produced. In both examples the first reaction is the chief one, and, under certain conditions, is the only one. Thus cuprous iodide furnishes iodo-derivatives only, with various diazo- chlorides and bromides, and cuprous chloride and diazobenzene iodide yield chlorobenzene, no iodobenzene being produced.

The third reaction, namely, the formation of azo-compounds, occurs when cuprous chloride, dissolved in hydrochloric acid, is added to the cold solution of the diazo-salt. Under these conditions, aniline, o-chloroaniline, and the o- and p-toluidines yield considerable quantities of azo-compound, but the nitroamines give diphenyl derivatives (see p. 61).

* Baeyer and Jaeger, *Ber.*, 1875, **8**, 893.
† *Ber.*, 1900, **33**, 2544.

CHAPTER VII

THE REACTIONS OF THE DIAZO-COMPOUNDS
(continued)

§ 1. Replacement of the diazo-group by cyanogen.—This is one of the most important of the diazo-decompositions, as it serves to introduce an additional carbon atom into the molecule; moreover, the nitriles formed in this way mostly yield the corresponding carboxylic acids.*

The preparation of p-toluonitrile is carried out as follows : 50 grams of copper sulphate are dissolved in 200 c.c. of water by heating on the water-bath, and a solution of 55 grams of potassium cyanide in 100 c.c. of water added gradually with continuous heating. Care must be taken to perform the operation under a hood as cyanogen is evolved. To this hot solution is now added during about ten minutes a diazo-solution prepared from 20 grams of p-toluidine, 50 grams of concentrated hydrochloric acid and 16 grams of sodium nitrite. The whole is now heated on the water-bath for a quarter of an hour and the toluonitrile distilled over with the steam. Here again care must be taken to get rid of the vapours as hydrogen cyanide is evolved. The nitrile distils as a yellow oil, which is purified by distillation.†

By treating a solution of diazobenzene chloride with potassium cyanide in the cold, a double compound of the diazo-cyanide and hydrogen cyanide, $C_6H_5.N_2.CN, HCN$, is formed.‡

The replacement is also effected by adding copper powder to a mixture of the diazo-salt and potassium cyanide, exactly as in the case of the preparation of the chloride.§

---

* Sometimes, however, these are formed with difficulty, owing probably to steric hindrance. Hofmann, *Ber.*, 1884, **17**, 1914; Küster and Stallberg, *Annalen*, 1894, **278**, 207 ; Cain, *Ber.*, 1895, **28**, 967.
† Gattermann, *Practical Methods of Organic Chemistry.*
‡ Gabriel, *Ber.*, 1879, **12**, 1637.
§ Gattermann, Hausknecht, Cantzler, and Ehrhardt, *Ber.*, 890, **23**, 1218.

§ 2. Replacement of the diazo-group by the cyano-group.—
This is effected by adding potassium cyanate to a diazo-sulphate and treating the mixture with copper powder, when the corresponding carbimide is obtained, thus—

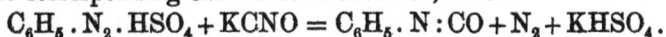

$$C_6H_5 . N_2 . HSO_4 + KCNO = C_6H_5 . N : CO + N_2 + KHSO_4 .$$

The potassium cyanate is prepared in the following manner : 100 grams of finely-powdered and sieved potassium ferro-cyanide are mixed with 75 grams of powdered potassium dichromate, each ingredient being first thoroughly dried. This mixture is added, in portions of 3–5 grams at a time, to an iron dish heated over a three-flame burner. The mass becomes black and is well stirred but should not be heated to the melting-point. On cooling it is extracted with five times its volume of 80 per cent. alcohol, and the cold solution stirred, when a white crystal powder of potassium cyanate separates, which is filtered and washed with small quantities of ether.

For the decomposition, 10 grams of aniline are dissolved in 100 grams of water and 20 grams of concentrated sulphuric acid, the solution cooled with ice, and diazotized with 7·5 grams of sodium nitrite. To the diazo-solution is added a concentrated aqueous solution of 9 grams of potassium cyanate and then 5 grams of copper powder, when evolution of nitrogen begins. A second 5 grams of copper powder is added and an oily layer of phenylcarbimide separates on the top of the liquid. This is skimmed off with a glass spoon, extracted with chloroform, and the chloroform solution filtered by the aid of the pump. More copper powder is added to the original solution until no more nitrogen is evolved and any further quantity of phenylcarbimide collected in the same way. The chloroform solutions are now separated from water, dried and freed from chloroform by evaporation. The residual oil, on distillation, yields pure phenylcarbimide.*

§ 3. Replacement of the diazo-group by the thiocyano-group.—This reaction is carried out by the aid of copper thiocyanate. For example : 31 grams of aniline are dissolved in 100 grams of concentrated sulphuric acid and 200 grams of water, and diazotized with 23 grams of sodium nitrite. To this

* *Ber.*, 1890, **23**, 1220; ibid., 1892, **25**, 1086.

E

solution is added a concentrated solution of 35 grams of potassium thiocyanate, and a paste of copper thiocyanate, obtained by dissolving 80 grams of copper sulphate and 150 grams of ferrous sulphate in water, precipitating with 35 grams of potassium thiocyanate and filtering. Nitrogen is evolved when this paste is added to the diazo-solution, and the reaction is complete after the whole has stood for three hours; the phenylthiocarbimide is then distilled with steam and rectified.*

§ 4. Replacement of the diazo-group by the group SK.—

When a diazo-sulphonate is warmed with an alcoholic solution of potassium sulphide nitrogen is evolved, and a thiophenol-sulphonic acid is formed.† Thus if the diazo-derivative of sulphanilic acid is used potassium p-thiophenolsulphonic acid results—

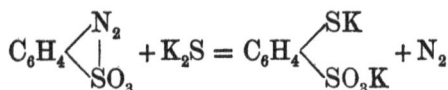

$$C_6H_4\!\!\begin{array}{c}N_2\\|\\SO_3\end{array} + K_2S = C_6H_4\!\!\begin{array}{c}SK\\\\SO_3K\end{array} + N_2$$

These thiophenols or mercaptans are also obtained by hydrolysing the xanthates produced by treating a diazo-salt with potassium xanthate.‡

By hydrolysing the xanthate from diazotized sulphanilic acid, in addition to the mercaptan, there is formed the ethosulphide—

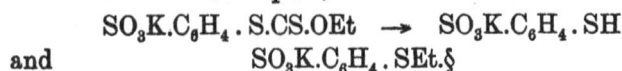

$$SO_3K.C_6H_4 . S.CS.OEt \;\rightarrow\; SO_3K.C_6H_4 . SH$$
and $$SO_3K.C_6H_4 . SEt.§$$

§ 5. Replacement of the diazo-group by sulphur.—

When hydrogen sulphide or ammonium sulphide acts on a solution of diazobenzene chloride or sulphate, the diazo-group is replaced by sulphur, and phenyl sulphide $(C_6H_5)_2S$ is produced ;‖ and if a solution of o-diazobenzoic acid sulphate is added to a cold saturated solution of sulphur dioxide in which copper powder is suspended, nitrogen is evolved, much copper passes into solution, part of the sulphur dioxide being oxidized to sulphuric acid, and the chief product is dithiosalicylic acid,

---

* *Ber.*, 1890, **23**, 738; compare Ibid., 770.
† Klason, *Ber.*, 1887, **20**, 349.
‡ Leuchart, *J. pr. Chem.*, 1890 [ii], **41**, 179.
§ Walter, *Proc.*, 1895, **11**, 141.
‖ Graebe and Mann, *Ber.*, 1882, **15**, 1683.

$(C_6H_4 . CO_2H)_2S$, which is obtained in a yield of about 50 per cent. of the theoretical.*

Another way in which sulphides are formed is by treating diazobenzene chloride with a colourless solution of copper sulphate (1 mol.) in sodium thiosulphate (6 mols.), that is, cuprous sodium thiosulphate. Phenyl sulphide is formed along with benzeneazodiphenyl.

Sulphanilic acid and o- and p-toluidine yield also the corresponding sulphides, but no diphenyl derivatives are produced. When a-naphthylamine is similarly treated there is no formation of sulphide, but a-azonaphthalene is obtained. †

§ 6. **Replacement of the diazo-group by the sulphonic acid group.**—Both the thiophenols and the disulphides yield the corresponding sulphonic acids on treatment with alkaline permanganate solution.‡

§ 7. **Replacement of the diazo-group by the nitro-group.**—This is brought about by treating a diazo-salt with nitrous acid and cuprous oxide. The amine is dissolved in two molecules of dilute nitric acid (hydrochloric acid is to be avoided), and, after being diazotized, a second molecule of sodium nitrite is added. The solution is then poured on finely-divided cuprous oxide and the reaction usually proceeds in the cold.

For example, the cuprous oxide is prepared by dissolving together 50 grams of copper sulphate and 15 grams of grape sugar in 100 grams of water. The solution is boiled and 20 grams of caustic soda, dissolved in 60 grams of water, added all at once. The mixture is neutralized with acetic acid.

On the other hand, 9 grams of aniline are dissolved in 50 grams of water and 20 grams of concentrated nitric acid (sp. gr. 1.4); 15 grams of sodium nitrite, dissolved in 50 grams of water, are added and then the diazo-solution is poured on the cuprous oxide gradually. When the reaction is finished the nitrobenzene is extracted by distillation with steam.§

The replacement also proceeds to a small extent in the

---

* Henderson, *Amer. Chem. J.*, 1899, **21**, 206.
† Börnstein, *Ber.*, 1901, **34**, 3968.
‡ F. Bayer & Co., D. R-P. 70286 of 1892; E. P. 11865 of 1892.
§ *Ber.*, 1887, **20**, 1495.

absence of cuprous oxide; thus when 2:4:6-tribromodiazo-benzene sulphate is treated with 20 molecular proportions of potassium nitrite, the corresponding 2:4:6-tribromo-1-nitrobenzene is formed, together with the quinonediazide—

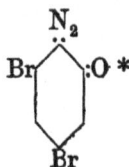

Nitrobenzene is also formed when diazobenzene perbromide is shaken with aqueous sodium hydroxide in the cold.†

The diazo-group of diazobenzene nitrate may also be replaced by the nitro-group by making use of the crystalline double salt of formula $Hg(NO_2)_2, 2C_6H_5 . N_2 . NO_3$, which is obtained by mixing solutions of diazobenzene nitrate and potassium mercuric nitrate. When this salt is boiled with water, it yields phenol and nitrophenol, but when treated with copper powder, a quantitative yield of nitrobenzene is produced.‡

Another method consists in mixing diazo-sulphates with a freshly prepared suspension of cupro-cupric sulphite and treating the mixture with excess of an alkali nitrite. By this means 2:4:6-tribromodiazobenzene sulphate gives a 65 per cent. yield of 2:4:6-tribromo-1-nitrobenzene, and β-diazo-naphthalene sulphate furnishes a 25 per cent. yield of β-nitro-naphthalene.§

§ 8. **Replacement of the diazo-group by the nitroso-group.** —This is effected by treating a diazobenzene chloride solution with an alkaline solution of potassium ferrocyanide.‖

§ 9. **Replacement of the diazo-group by the amino-group.** —This replacement is effected by adding hydroxylamine to

---

* Orton, *Trans.*, 1903, **83**, 806.
† Bamberger, *Ber.*, 1894, **27**, 1273.
‡ Hantzsch and Blagden, *Ber.*, 1900, **33**, 2544.
§ Hantzsch and Blagden, loc. cit.
‖ Bamberger and Storch, *Ber.*, 1893, **26**, 471.

a solution of a diazo-salt;[*] thus, aniline may be obtained from diazobenzene chloride, and p-toluidine from its diazo-salt. An interesting example occurs in the anthracene series. When the anhydride of 1-diazoanthraquinone-2-sulphonic acid

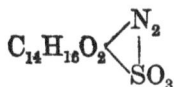

$$C_{14}H_{16}O_2 \left\langle \begin{array}{c} N_2 \\ | \\ SO_3 \end{array} \right.$$

is suspended in water and treated with ammonia or ammonium carbonate, nitrogen is evolved, and the original aminosulphonic acid is obtained. Further, when this diazo-compound is treated with hydroxylamine, a diazohydroxyamide

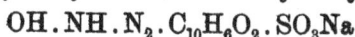

$$OH . NH . N_2 . C_{10}H_6O_2 . SO_3Na$$

is formed, which is transformed by concentrated sulphuric acid into 1-amino-4-hydroxyanthraquinone-2-sulphonic acid.

A similar reaction takes place when hydrazine is substituted for hydroxylamine; in this way, both the amino- and hydroxyl-groups are introduced into the molecule when the diazo-group is eliminated.[†]

§ 10. **Replacement of the diazo-group by the acetoxy-group.**—Meldola and East [‡] found that when certain azo-derivatives of β-naphthylamine, containing an amino-group, are diazotized in warm glacial acetic acid, the diazo-group is replaced by the acetoxy-group, and Orndorff [§] has shown that this reaction may be applied generally for the preparation of aromatic acetates.

The acetoxy-group may be readily converted into the hydroxy-group by hydrolysis, so that this method is useful in effecting the replacement of the diazo- by the hydroxy-group in such cases where the normal decomposition with water does not take place.

* Mai, *Ber.*, 1892, **25**, 372.   † Wacker, *Ber.*, 1902, **35**, 2593, 3920.
‡ *Trans.*, 1888, **53**, 460.   § *Amer. Chem. J.*, 1888, **10**, 368.

CHAPTER VIII

ACTION OF VARIOUS REAGENTS ON DIAZO-COMPOUNDS

§ 1. **Sulphur dioxide.**—When diazobenzene chloride is treated with sulphur dioxide in aqueous solution in the cold, reduction takes place with formation of a hydrazine, and at the same time a second reaction proceeds by which the nitrogen is eliminated and the sulphonic acid group takes its place. These two products condense together in the nascent state and a sulphazide is formed[*]—

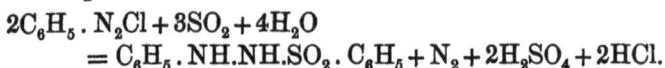

$$2C_6H_5 . N_2Cl + 3SO_2 + 4H_2O$$
$$= C_6H_5 . NH.NH.SO_2 . C_6H_5 + N_2 + 2H_2SO_4 + 2HCl.$$

These compounds are also formed by dissolving the amine in 95 per cent. alcohol, saturating the solution with sulphur dioxide, and adding a concentrated aqueous solution of potassium nitrite.[†]

A differently constituted product results when the neutral diazo-compound, prepared from p-nitroaniline (p-nitrodiazobenzene hydroxide, $NO_2 . C_6H_4 . N_2 . OH$), is dissolved in absolute alcohol and subjected to the action of dry sulphur dioxide at 0°–5°. p-Nitrobenzenediazo-p-nitrobenzenesulphone

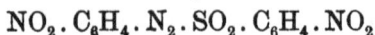

$$NO_2 . C_6H_4 . N_2 . SO_2 . C_6H_4 . NO_2$$

is formed.[‡]

These sulphones are also obtained by treating a diazo-salt with benzenesulphinic acid.[§]

Condensation products in which two benzene nuclei exist are also obtained when a diazo-salt is subjected to the action of sulphur dioxide in presence of a not too large excess of

---

[*] Koenigs, *Ber.*, 1877, **10**, 1531.   [†] Ulatowski, *Ber.*, 1887, **20**, 1238.
[‡] Ekbom, *Ber.*, 1902, **35**, 656.
[§] Hantzsch and Singer, *Ber.*, 1897, **30**, 312.

sulphuric acid.* Diazobenzene chloride, under these conditions, yields a compound of formula

$$C_6H_5 . N : N.C_6H_4 . NH.NH.SO_3H,$$

and *m*-diazotoluene chloride a compound of analogous constitution. The reaction takes a different course when sulphites are employed. With neutral alkali sulphites the corresponding diazo-salts are obtained—

$$C_6H_5 . N_2Cl + K_2SO_3 = C_6H_5 . N_2 . SO_3K + KCl.†$$

Acid sulphites furnish hydrazinesulphonic acids of formula $C_6H_5 . NH.NH.SO_3X.$‡ Also when a solution of sodium hyposulphite is allowed to react with diazobenzene sulphate or chloride, the chief product is sodium phenylhydrazine-β-sulphonate. There are also formed small quantities of diazobenzeneimide and benzenesulphonphenylhydrazine.§ In alkaline solution, the diazo-group is replaced by hydrogen (see p. 42).

§ 2. **Replacement of the diazo-group by the sulphinic acid group.**—The formation of sulphinic acids by the direct action of sulphurous acid on diazo-salts was first observed by Müller and Wiesinger,‖ but the replacement is best carried out by Gattermann's method, using copper powder.¶ A solution of the diazo-sulphate, containing an excess of sulphuric acid, is saturated with sulphur dioxide, the solution being kept cold. Each 100 c.c. of the solution should absorb about 15 grams of the gas. Copper powder is now added gradually to the solution (which should be clear), ice being added to keep the solution cold during the operation. The addition of copper is continued with vigorous stirring until no more nitrogen is evolved. As some sulphur dioxide is carried off

* 15 grams of aniline and 50 grams of concentrated sulphuric acid. Tröger, Hille, and Vesterling, *J. pr. Chem.*, 1905 [ii], **72**, 511 ; Tröger and Schaub, *Arch. Pharm.*, 1906, **244**, 302 ; Tröger and Franke, ibid., 307 ; Tröger, Warnecke, and Schaub, ibid., 312 ; Tröger, Berlin, and Franke, ibid., 326.
† Griess, *Ber.*, 1876, **9**, 1653.
‡ Schmitt and Glutz, *Ber.*, 1869, **2**, 51 ; Strecker and Römer, *Ber.*, 1871, **4**, 784 ; E. Fischer, *Ber.*, 1875, **8**, 589.
§ Grandmougin, *Ber.*, 1907, **40**, 422.   ‖ *Ber.*, 1879, **12**, 1348.
¶ *Ber.*, 1899, **32**, 1136.

with the nitrogen, a further quantity is passed through the mixture during the reaction. The sulphinic acid is extracted from the product by means of ether. In the case of the diazotized naphthylamines it is better to add the diazo-solution to a mixture of copper powder and a saturated solution of sulphurous acid.

§ 3. Hydrogen sulphide.—When hydrogen sulphide is passed through an aqueous, nearly neutral, solution of p-nitro-diazobenzene chloride at 0°, the diazo-sulphide

$$(NO_2 . C_6H_4 . N_2)_2S$$

is produced.

In hydrochloric acid solution, the mercaptan hydrosulphide, $NO_2 . C_6H_4 . N_2SH.H_2S$, is first formed, and on prolonging the passage of the gas, the disulphide, $(NO_2 . C_6H_4 . N_2)_2S_2$, results.*

On warming diazotized sulphanilic acid with alcoholic potassium sulphide, the diazo-nitrogen is expelled and the dipotassium salt of p-thiophenolsulphonic acid is produced—

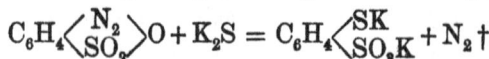

$$C_6H_4 {\Large\langle} {N_2 \atop SO_2} {\Large\rangle} O + K_2S = C_6H_4 {\Large\langle} {SK \atop SO_3K} + N_2 \dagger$$

and mercaptan combines with diazo-salts to form an intermediate compound which loses nitrogen on warming—

$$C_6H_4 {\Large\langle} {N_2 \atop SO_2} {\Large\rangle} O \ \rightarrow \ C_6H_4 {\Large\langle} {N_2.S.C_2H_5 \atop SO_3H} \ \rightarrow \ C_6H_4 {\Large\langle} {S.C_2H_5 \atop SO_3H}$$

Similarly phenyl mercaptan forms corresponding thiophenol ethers.‡

§ 4. Replacement of the diazo-group by the azoimino-group.

1. **Action of ammonia.**—Griess examined the action of concentrated aqueous ammonia on diazobenzene nitrate,§ and obtained an extremely unstable compound which decomposed into phenol, aniline, and nitrogen.

This substance was shown by von Pechmann ‖ to consist of bisdiazobenzeneamide, the reaction proceeding as follows—

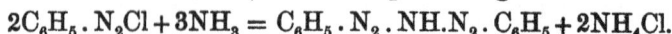

$$2C_6H_5 . N_2Cl + 3NH_3 = C_6H_5 . N_2 . NH.N_2 . C_6H_5 + 2NH_4Cl.$$

* Bamberger and Kraus, *Ber.*, 1896, **29**, 272.
† *Ber.*, 1887, **20**, 350.   ‡ Hantzsch and Freese, *Ber.*, 1895, **28**, 3237.
§ *Annalen*, 1866, **137**, 81.
‖ *Ber.*, 1894, **27**, 898; ibid., 1895, **28**, 171.

A similar substance is obtained from p-diazotoluene chloride, but p-nitrodiazobenzene chloride yields only p-dinitrodiazoaminobenzene under the same conditions.

On extending this reaction to diazobenzene perbromide, Griess obtained the first of a very important series of new compounds, namely, the diazoimides, containing three atoms of nitrogen united together. The empirical formula is $C_6H_5N_3$, and Kekulé proposed for it the constitutional formula—

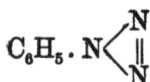

$$C_6H_5 . N\diagdown\!\!\begin{array}{c} N \\ \| \\ N \end{array}$$

$$C_6H_5 . NBr.NBr_2 + NH_3 = C_6H_5 . N_3 + 3HBr.$$

Diazobenzeneimide or phenylazoimide is a yellow oil possessing a stupefying odour. It boils at 59° under a pressure of 12 mm. and explodes when heated at the ordinary pressure. When it is heated with hydrochloric acid nitrogen is evolved and chlorobenzene is obtained*; with sulphuric acid two-thirds of the nitrogen is eliminated and aminophenol is produced†. Diazobenzeneimide is also obtained when hydroxylamine acts on diazobenzene sulphate,‡

$$C_6H_5 . N_2 . HSO_4 + NH_2 . OH = C_6H_5 . N_3 + H_2O + H_2SO_4$$

and by the elimination of water from nitrosophenylhydrazine—

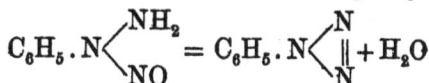

$$C_6H_5 . N\diagdown\!\!\begin{array}{c} NH_2 \\ NO \end{array} = C_6H_5 . N\diagdown\!\!\begin{array}{c} N \\ \| \\ N \end{array} + H_2O$$

2. **Action of hydrazine.**—The action of diazo-salts on phenylhydrazine was first studied by Griess,§ who obtained diazobenzeneimide by treating phenylhydrazine with m-diazobenzoic acid—

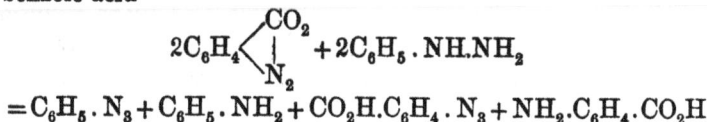

$$2C_6H_4\diagdown\!\!\begin{array}{c} CO_2 \\ | \\ N_2 \end{array} + 2C_6H_5 . NH.NH_2$$

$$= C_6H_5 . N_3 + C_6H_5 . NH_2 + CO_2H.C_6H_4 . N_3 + NH_2.C_6H_4.CO_2H$$

* *Ber.*, 1886, **19**, 313.   † *Ber.*, 1894, **27**, 192.
‡ *Ber.*, 1892, **25**, 372; 1893, **26**, 1271; compare also Forster and Fierz, *Trans.*, 1907, **91**, 855, 1350.
§ *Ber.*, 1876, **9**, 1659.

and also by the interaction of diazobenzene and m-hydrazino-benzoic acid—

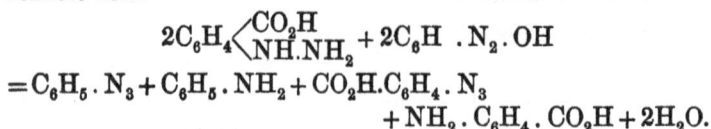

$$2C_6H_4{\begin{smallmatrix}CO_2H\\NH.NH_2\end{smallmatrix}} + 2C_6H . N_2 . OH$$
$$= C_6H_5 . N_3 + C_6H_5 . NH_2 + CO_2H.C_6H_4 . N_3$$
$$+ NH_2 . C_6H_4 . CO_2H + 2H_2O.$$

The same compound was obtained by E. Fischer by acting on phenylhydrazine with diazobenzene sulphate—

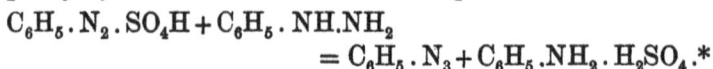

$$C_6H_5 . N_2 . SO_4H + C_6H_5 . NH.NH_2$$
$$= C_6H_5 . N_3 + C_6H_5 .NH_2 . H_2SO_4.*$$

It has been shown, however, that in addition to the above compound a substance is formed of formula

$$C_6H_5 . N_2 . N(C_6H_5) . NH_2,$$

to which the name diazobenzenephenylhydrazide is given.[†] When this compound is oxidized with dilute permanganate solution, bisdiazobenzenediphenyltetrazone is obtained. This compound, of formula

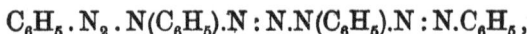

$$C_6H_5 . N_2 . N(C_6H_5).N : N.N(C_6H_5).N : N.C_6H_5,$$

contains a chain of no less than eight nitrogen atoms.

When hydrazine itself is substituted for the phenyl derivative two reactions proceed: on the one hand, we have the formation of diazobenzeneimide and ammonia, and on the other, aniline and azoimide are produced, thus—

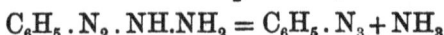

$$C_6H_5 . N_2 . NH.NH_2 = C_6H_5 . N_3 + NH_3$$
and $$C_6H_5 . N_2 . NH.NH_2 = C_6H_5 . NH_2 + N_3H ; ‡$$

the latter reaction, however, proceeds to only a slight extent.

3. Action of azoimide.—The azoimides are also obtained by adding a solution of azoimide or its sodium salt to a diazo-solution containing excess of sulphuric acid. The resulting azoimide is extracted with ether.§ (For the preparation of azoimide by reactions which do not involve the use of diazo-

---

* *Ber.*, 1877, **10**, 1334 ; *Annalen*, 1878, **190**, 94 ; compare also Griess, *Ber.*, 1887, **20**, 1528 ; ibid., 1888, **21**, 3415.
† Wohl and Schiff, *Ber.*, 1900, **33**, 2741.
‡ *Ber.*, 1893, **26**, 88, 1263.
§ Noelting and Michael, *Ber.*, 1893, **26**, 86.

compounds, textbooks on Organic Chemistry should be consulted.)

§ 5. **Benzoyl chloride.**—When diazo-salts are treated with an aqueous suspension of benzoyl chloride and copper powder, dibenzoylhydrazines, $RN(CO.C_6H_5)N(CO.C_6H_5)R$, are obtained.*

* Biehringer and Busch, *Ber.*, 1902, **35**, 1964.

CHAPTER IX

FORMATION OF DIPHENYL DERIVATIVES IN THE DIAZO-REACTION

By acting on diazobenzene nitrate with potassium ferrocyanide, Griess * obtained azobenzene, a substance having the formula $C_{18}H_{14}N_2$, and a brownish oil. The second of these was shown later to be benzeneazodiphenyl—

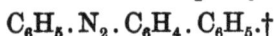

$$C_6H_5 . N_2 . C_6H_4 . C_6H_5.†$$

Griess also observed the formation of p-diphenol by the decomposition of the double salt, $(C_6H_5 . N_2Cl)_2$, $SnCl_4$. ‡

As will be seen later (p. 73) diazo-salts combine with amines to form diazoamino-compounds, and these pass, by molecular change, into aminoazo-compounds. In the simplest case, that of diazobenzene chloride and aniline, in addition to aminoazobenzene, o- and p-aminodiphenyl are formed.§

A similar case has not been observed, however, when a diazosalt acts on a phenol, but when nitrosophenol is thus treated, diphenyl derivatives are largely produced.‖

Also in the preparation of phenol from diazobenzene sulphate o- and p-hydroxydiphenyl are formed. ¶

It is evident, therefore, that at the moment when the diazonitrogen separates from the benzene nucleus, two of the latter unite at this point.

Diphenyl may be prepared in good yield by Gattermann's method of adding copper powder to a solution of diazobenzene sulphate in alcohol.** 31 grams of aniline are dissolved in

* *Annalen*, 1866, **137**, 39; *Ber.*, 1876, **9**, 132.
† Locher, *Ber.*, 1888, **21**, 911; compare also p. 62.
‡ *Ber.*, 1885, **18**, 960.
§ Hirsch, *Ber.*, 1892, **25**, 1973; see also Heusler, *Annalen*, 1890, **260**, 227.
‖ Borsche, *Ber.*, 1899, **32**, 2935; *Annalen*, 1900, **312**, 211.
¶ Hirsch, *Ber.*, 1890, **23**, 3705; *J. pr. Chem.*, 1885 [ii], **32**, 117; compare also Norris, Macintyre, and Corse, *Amer. Chem. J.*, 1903, **29**, 120.
** *Ber.*, 1890, **23**, 1226.

40 grams of concentrated sulphuric acid and 150 grams of water, and the solution diazotized in the usual manner with 23 grams of sodium nitrite. 100 grams of alcohol (90 per cent.) are now added, and then 50 grams of copper powder. Nitrogen is evolved and the temperature rises to about 30-40°. After about one hour, when the reaction is finished, the whole is distilled with steam; alcohol passes over, and when the distillate gives, on addition of water, a solid substance, the receiver is changed, and crystals of diphenyl are collected. Instead of copper powder, 100 grams of zinc dust or iron powder may be used.

Diphenyl is also obtained by the action of stannous chloride on diazobenzene chloride or formate. *

A large number of similar condensation products are obtained by subjecting mixtures of diazo-salts and hydrocarbons or similar ring-compounds to the action of aluminium chloride; thus diphenyl is obtained from diazobenzene chloride and benzene, and the corresponding phenyl derivative results from the condensation of this diazo-salt with thiophen, pyridine, and quinoline. †

In applying the cuprous chloride and copper powder methods for the production of chloro-derivatives to the case of many nitrodiazo- and chloronitrodiazo-salts, a remarkable tendency towards the formation of diphenyl derivatives has been observed.

Thus, when o-nitrodiazobenzene chloride is acted on by copper powder, a yield of 60 per cent. of 2 : 2′-dinitrodiphenyl is obtained;‡ by using cuprous chloride a yield of 68 per cent. was observed. §

The diphenyl reaction is also brought about by treating diazo-salts with cuprous oxide dissolved in ammonia,‖ and corresponding derivatives are formed by treating diazo-salts with zinc ethyl.¶

* Culmann and Gasiorowski, *J. pr. Chem.*, 1889 [ii], **40**, 97.
† Möhlau and Berger, *Ber.*, 1893, **26**, 1994; see also Kühling, *Ber.*, 1895, **28**, 41; 1896, **29**, 165, and Bamberger, *Ber.*, 1895, **28**, 403.
‡ Niementowski, *Ber.*, 1901, **34**, 3325.
§ Ullmann and Forgan, *Ber.*, 1901, **34**, 3802; D. R-P. 126961.
‖ Vorländer and F. Meyer, *Annalen*, 1902, **320**, 122.
¶ Bamberger and Tichwinsky, *Ber.*, 1902, **35**, 4179. Tichwinsky, *J. Russ. Phys. Chem. Soc.*, 1903, **35**, 155, 675; 1904, **36**, 1052.

A similar condensation takes place in the naphthalene series; thus when β-diazonaphthalene sulphate is dissolved in alcohol and treated with zinc dust, to which has been added a very little powdered copper sulphate, $\beta\beta$-dinaphthyl is formed,* and when a cold neutral solution of diazobenzene chloride is mixed with a solution containing 1 molecular proportion of copper sulphate and 6 molecular proportions of sodium thiosulphate (a solution of the salt $Cu_2S_2O_3$, $3\,Na_2S_2O_3$, $6\,H_2O$), benzene-azodiphenyl, $C_6H_5 . N_2 . C_6H_4 . C_6H_5$, is produced, together with phenyl sulphide, $(C_6H_5)_2S$.†

* Chattaway, *Trans.*, 1895, **67**, 653.
† Börnstein, *Ber.*, 1901, **34**, 3968.

CHAPTER X

INTERCHANGE OF GROUPS IN DIAZO-COMPOUNDS

A CURIOUS reaction was noticed by Meldola,* who found that when 3 : 4-dinitro-o-anisidine

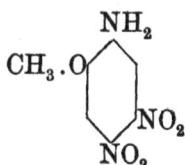

is diazotized in acetic acid solution, the resulting diazo-compound only contains one nitro-group; the other having been eliminated during diazotization, a substance of formula

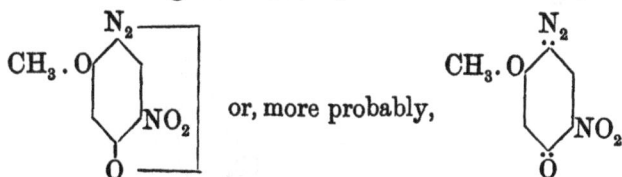

 or, more probably,

being obtained.† The nitro-group has thus been replaced by hydroxyl in the process, being itself liberated in the form of nitrous acid.

In a similar manner dinitro-p-anisidine

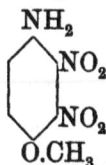

on being diazotized in presence of acetic acid ‡ loses a nitro-group, the diazo-compound formed giving with β-naphthol a substance of formula

* *Trans.*, 1900, **77**, 1172. † *Proc.*, 1901, **17**, 135.
‡ Meldola and Eyre, *Trans.*, 1902, **81**, 988.

$$N_2 . C_{10}H_6 . OH$$

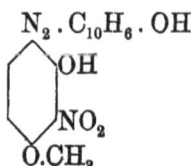

the diazo-compound itself not having been isolated.

In nitric or sulphuric acid solution the nitro-group remains unaffected, but in presence of hydrochloric acid the nitro-group adjacent to the diazo-group is replaced by chlorine.

Meldola and his pupils have found that when a nitro-group is in the ortho- or para-position with respect to an amino-group, no displacement of the nitro-group takes place on diazotization unless there is a second nitro-group adjacent to the first (mobile) group.

It has further been observed that when a methoxy-group is in the para-position with respect to the amino-group, and at the same time has a nitro-group in an adjacent position, demethylation takes place on diazotization.*

Thus the compounds

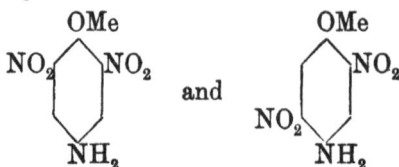

 and

yield the corresponding quinonediazides of dinitrobenzene—

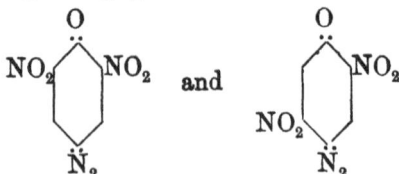

 and

Also when m-phenylenediaminedisulphonic acid is tetrazotized, a sulphonic acid group is replaced by hydroxyl with formation of tetrazophenolsulphonic acid.†

In some other cases which have been observed, it has been possible to obtain a nitrodiazo-compound which, even on dilu-

* Meldola and Stephens, Trans., 1905, 87, 1205.
† E. P. 18283 of 1903.

tion with water, soon loses a nitro-group. Thus if the dinitro-
β-naphthylamine of formula

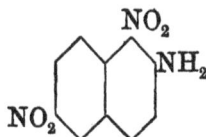

$$NO_2$$
$$NH_2$$
$$NO_2$$

is diazotized in concentrated sulphuric acid solution and
poured into ice-cold water, a precipitate is formed after a short
time consisting of a diazo-oxide, to which is assigned the
formula

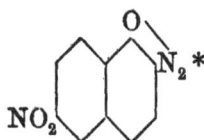

$$O$$
$$N_2 *$$
$$NO_2$$

The nitro-group in the α-position is thus replaced by
hydroxyl. Similarly from the dinitronaphthylamine

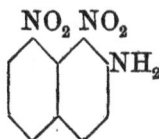

$$NO_2 NO_2$$
$$NH_2$$

the corresponding mononitrodiazo-oxide

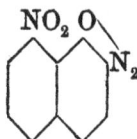

$$NO_2 O$$
$$N_2$$

is obtained. The formation of the nitrodiazo-oxide

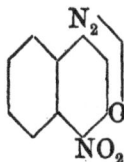

$$N_2$$
$$O$$
$$NO_2$$

* Gaess and Ammelburg, *Ber.*, 1894, **27**, 2211.

F

from the corresponding dinitro-α-naphthylamine

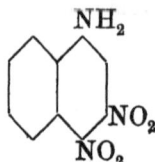

takes place in exactly the same way.*

In all the above cases it will have been noted that the nitro-group which is eliminated reappears as free nitrous acid. This has led Meldola and Eyre † to make the experiment of starting the diazotization of the above-mentioned dinitro-*o*-anisidine with a small quantity of nitrous acid (one quarter of the theoretical amount was used); and they observed that the diazotization was continued by the nitrous acid thus eliminated.

This transformation is not confined to those diazo-compounds containing only nitro-groups. Many other cases are known; thus Meldola and Streatfeild ‡ found that when the sulphate of dibromo-β-naphthylamine

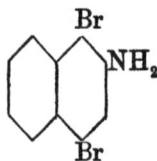

was diazotized in presence of acetic acid and the resulting mixture raised to the boiling-point, the normal reaction, namely, replacement of the diazo-group by hydroxyl (see p. 29) did not take place, but bromine was displaced and a diazo-oxide was formed—

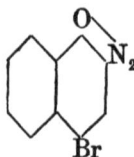

In a similar manner chlorobromo-β-naphthylamine yielded a bromodiazo-oxide—

---

* Friedländer, *Ber.*, 1895, **28**, 1951.
† *Trans.*, 1901, **79**, 1076.   ‡ *Trans.*, 1895, **67**, 908.

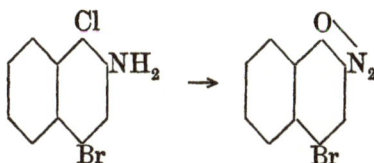

Similar substitutions of a halogen-group by hydroxyl have been observed to take place by merely treating the diazo-salt with alkalis. Thus $2:4:6$-tribromodiazobenzene chloride yields the dibromodiazo-oxide

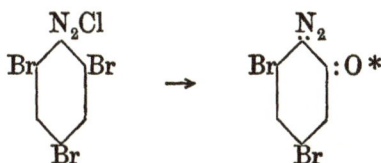

The same reaction takes place in the case of 2-chloro-3-nitroaniline-5-sulphonic acid,

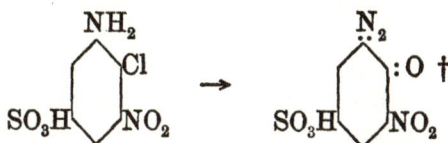

and the tetrazo-derivative of 2-chloro-m-phenylenediamine 5-sulphonic acid, under the same conditions, loses chlorine, a hydroxyl-group taking its place—

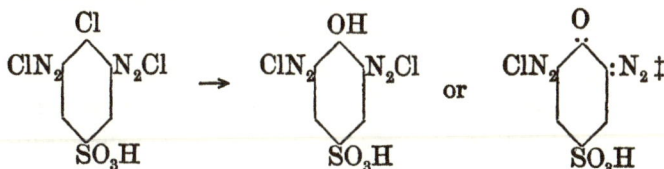

A sulphonic acid group also undergoes this change; thus the compound

---

* Bamberger and Kraus, *Vierteljahrssch. Ges. Zürich*, 1899, **24**, 257; *Ber.*, 1906, **39**, 4248; Bamberger, *Annalen*, 1899, **305**, 289. Compare also Silberstein, *J. pr. Chem.* 1883 [ii], **27**, 98.

† Badische Anilin- und Soda-Fabrik, D. R-P. 141750.

‡ E. P. 16811 of 1901.

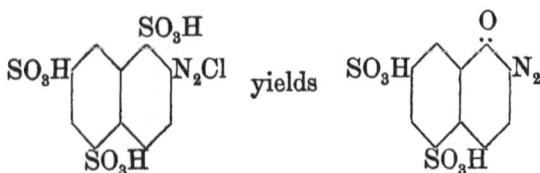

when diazotized and rendered alkaline.*

The replacement proceeds even when the diazo-salt of a weak acid such as the acetate, carbonate, bicarbonate, oxalate, &c., is allowed to stand ; this takes place in the case of 2 : 5 : 6-trichloroaniline-m-sulphonic acid, o-nitroaniline-p-sulphonic acid, and 2 : 4-dinitroaniline.†

The transformation of 2 : 4 : 6-tribromo- and trichloro-diazo-benzene takes the same course as shown above,‡ and a similar phenomenon occurs in the case of a considerable number of halogen-derivatives of the benzene and naphthalene series.§

Some striking molecular transformations have been observed by Hantzsch.‖ If p-chlorodiazobenzene thiocyanate (prepared by adding potassium thiocyanate to the diazo-chloride) is dissolved in alcohol containing a trace of hydrochloric acid, the thiocyano-group changes place with the chlorine atom, and on adding ether to the solution, p-thiocyanodiazobenzene chloride is precipitated, thus—

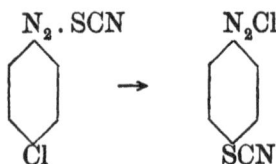

Similarly, many brominated diazo-chlorides pass into chlorinated diazo-bromides ; ¶ for example, 2 : 4 : 6-tribromodiazo-benzene chloride is converted into chlorodibromodiazobenzene bromide.**

* E. P. 23993 of 1902.   † E. P. 20551 of 1901.
‡ Orton, *Proc. Roy. Soc.*, 1902, **71**, 153.
§ Orton, *Proc.*, 1902, **18**, 252 ; *Trans.*, 1903, **83**, 796 ; 1907, **91**, 1554 ; Badische Anilin- und Soda-Fabrik, E. Ps. 1561, 6615 of 1902 ; 16995, 27372 of 1903 ; 4997 of 1904 ; Noelting and Battegay, *Ber.*, 1906, **39**, 79.
‖ *Ber.*, 1896, **29**, 947.
¶ Hantzsch, Schleissing, and Jäger, *Ber.*, 1897, **30**, 2334 ; see also *Ber.*, 1898, 31, 1253.
** Hantzsch and Smythe, *Ber.*, 1900, **33**, 505.

This transformation has been studied quantitatively, and has been found to proceed according to the following laws :—

(1) The bromine atoms are replaced only when present in the para- or ortho-position with respect to the diazo-group, those in the ortho-position being most readily removed. A bromine atom in the meta-position is not affected.

(2) The ease of transformation increases with the number of bromine atoms present.

(3) The transformation constant, calculated from the equation for a unimolecular reaction,

$$k = \frac{1}{t} \log \frac{A}{A-x},$$

increases with the temperature and is also influenced by the solvent, having its minimum value in water, and becoming greater as the series of alcohols is ascended.

(4) The diazo-salts containing two bromine atoms are stable when dry, but are rapidly transformed in ethyl alcohol ; 2 : 4 : 6-tribromodiazobenzene chloride becomes transformed even in the dry state.

A corresponding isomeric change does not take place in the case of tri-iododiazobenzene chloride or tribromodiazobenzene fluoride.*

Lastly, a remarkable change is undergone by 1-nitrodiazo-β-naphthalene chloride which is transformed in presence of glacial acetic acid into 1-chlorodiazo-β-naphthalene nitrite.†

* Hantzsch, *Ber.*, 1903, **36**, 2069.
† Morgan, *Trans.*, 1902, **81**, 1376.

CHAPTER XI

ACTION OF LIGHT ON DIAZO-COMPOUNDS

MOST investigators who have worked with diazo-compounds have noticed that they are very easily changed by the action of light. Thus Berthelot and Vielle in 1881 * recorded the observation that when diazobenzene nitrate was exposed to light it became rose-coloured.

This decomposition has been made the basis of photographic processes; thus Feer in 1889 † exposed a film coated with a mixture of a diazo-sulphite and a phenol or amine to light.

A decomposition of the former occurred which was followed by the formation of an azo-compound, and hence the production of a coloured negative.

Green, Cross, and Bevan ‡ coated films with diazotized primuline, the decomposition of which was proportional to the intensity of the light; this formed the 'negative', and a 'positive' was developed by treatment with an amine or a phenol. Those parts of the negative which had been exposed to bright light gave no colour with the component, owing to the destruction of the diazo-compound with evolution of nitrogen and formation of a phenol. They concluded, however, that union of the diazo-compound with the medium (cellulose) was necessary, for the free diazo-primuline when exposed to light in a thin film was either not decomposed at all or only after long exposure.

Andresen § examined the behaviour of the diazo-salts of the two naphthylamines, and showed that the reaction was similar to that effected by heat, namely, that phenols were formed thus—

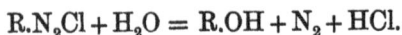

$$R.N_2Cl + H_2O = R.OH + N_2 + HCl.$$

Ruff and Stein ‖ arrived at the following conclusions with

* *Compt. rend.*, 1881, **92**, 1074.   † D. R-P. 53455.
‡ D.R-P. 56606, *Ber.*, 1890, **23**, 3131 ; *J. Soc. Chem. Ind.*, 1890, **9**, 1001.
§ *Photographische Correspondenz*, 1895.
‖ *Ber.*, 1901, **34**, 1668.

regard to the action of light on substituted diazobenzene chlorides.

Those which contain a negative group (OH, NO_2, CO_2H) in the para-position are more sensitive than those containing a similarly situated positive group (Cl, CH_3); the influence of the nitro-group is greatest. Ortho- and para-substituted groups have about the same effect, either in increasing or decreasing sensitiveness; this effect is always less than that of a meta-group. In the case of diazo-salts derived from different nuclei, the sensitiveness to light increases with the number of atoms in the nucleus; thus the diazo-salt from 3-aminocarbazole is nearly five times as sensitive as that from p-toluidine.

As regards the decomposition of diazo- and tetrazo-compounds an equal number of diazo-groups are destroyed by light in the same time; thus the same number of minutes is necessary to decompose completely the diazo-salt from an $N/10$ solution of p-aminodiphenyl as from an $N/20$ solution of benzidine.

Orton, Coates, and Burdett * were the first to investigate extensively this reaction. Solutions of the diazo-salts of various aromatic amines were found to decompose under the action of light in the manner indicated by Andresen, but the mechanism whereby the phenolic decomposition is effected must be very different from that induced by the action of heat, for the remarkable fact was discovered that many of the diazo-compounds which are decomposed by water or acids only with very great difficulty, and then only to a very slight extent, for example, the diazo-salt of 2:4:6-tribromoaniline, undergo rapid transformation under the action of light with quantitative formation of the corresponding phenol.

A similar instance of this difference in stability towards heat and light had been noticed by Meldola, Woolcott, and Wray,† who found that the compound

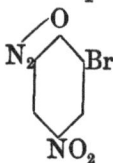

* *Trans.*, 1907, **91**, 35.   † *Trans.*, 1896, **69**, 1327.

was stable towards boiling water, but that it decomposed gradually and became brown on exposure to light.

The *syn*-diazo-cyanide of 2 : 4 : 6-tribromoaniline in benzene solution changes under the action of light into the corresponding *anti*-compound.*

* Ciusa, *Atti. R. Accad. Lincei*, 1906 [v], **15**, ii, 136 ; for an explanation of the terms *syn* and *anti* see p. 123.

CHAPTER XII

DIAZOAMINO-COMPOUNDS

THE diazoamino-compounds are formed by the condensation of a diazo-salt with primary or secondary amines in presence of sodium acetate, thus—

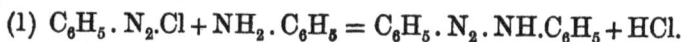

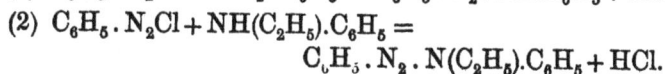

(1) $C_6H_5 . N_2.Cl + NH_2 . C_6H_5 = C_6H_5 . N_2 . NH.C_6H_5 + HCl.$

(2) $C_6H_5 . N_2Cl + NH(C_2H_5).C_6H_5 =$
$$C_6H_5 . N_2 . N(C_2H_5).C_6H_5 + HCl.$$

The preparation of diazoaminobenzene is carried out as follows: 10 grams of aniline are dissolved in 100 c.c. of water and concentrated hydrochloric acid corresponding to 12 grams HCl. The solution is diazotized by adding a solution of 8 grams of sodium nitrite with the usual precautions. On the other hand, 10 grams of aniline are dissolved in 50 grams of water and exactly the theoretical quantity of hydrochloric acid. After cooling this solution with ice it is added to the diazo-solution, and then, immediately, a cold concentrated solution of 50 grams of sodium acetate. After standing for half an hour the diazoaminobenzene is filtered off, washed with water, dried on a porous plate, and crystallized from light petroleum.

When aromatic diazo-compounds are allowed to act on aliphatic amines, similar diazoamino-compounds are obtained. Thus methylamine and ethylamine yield with diazobenzene, diazobenzenemethylamide (phenylmethyltriazen),

$$C_6H_5 . N_2 . NH.CH_3,$$

and diazobenzene-ethylamide (phenylethyltriazen),

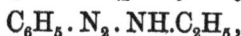

$$C_6H_5 . N_2 . NH.C_2H_5,$$

respectively.*

These compounds possess the formulae assigned to them, and are not tautomeric. †

* Dimroth, Ber., 1903, 36, 909; 1905, 38, 670, 2328; compare also Goldschmidt and Holm., Ber., 1888, 21, 1016.
† Dimroth, Eble, and Gruhl, Ber., 1907, 40, 2390.

The formation takes place also when an alkali nitrite is added to a solution of an amine containing no free mineral acid—

$$2C_6H_5 . NH_2 . HCl + NaNO_2$$
$$= C_6H_5 . N_2 . NH.C_6H_5 + NaCl + HCl + 2H_2O.$$

If two molecules of a diazo-salt condense with one of a primary amine, a bisdiazoamino-compound is formed—

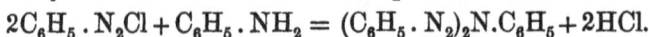

$$2C_6H_5 . N_2Cl + C_6H_5 . NH_2 = (C_6H_5 . N_2)_2N.C_6H_5 + 2HCl.$$

A modification of this method is to allow a molecule of a diazo-salt to act on a molecule of a diazoamino-compound—

$$C_6H_5 . N_2Cl + C_6H_5 . N_2 . NH.C_6H_5 = (C_6H_5.N_2)_2N.C_6H_5 + HCl.*$$

The primary monoamines of the benzene series all yield diazoamines, those containing the groups Cl, NO_2, CN, &c., most readily, but the monoalkylated monoamines of this series show a tendency to form azo-compounds; for example, methylaniline, when treated with diazobenzenesulphonic acid, yields a mixture of the diazoamino-compound,

$$SO_3H.C_6H_4 . N_2 . N(CH_3) . C_6H_5,$$

and the isomeric aminoazo-compound,

$$SO_3H.C_6H_4 . N_2 . C_6H_4 . NH.CH_3 . †$$

A number of bases, as for example diphenylamine, the naphthylamines and their monoalkyl-derivatives, m-phenylenediamine and certain of its homologues and substitution products, form aminoazo-compounds direct. Dimethylaniline and some other tertiary amines also yield aminoazo-compounds; here, of course, no diazoamine can be formed. Griess discovered the remarkable fact that the same compound is obtained from, for example, diazobenzene chloride and p-toluidine on the one hand, and p-diazotoluene chloride and aniline on the other.

According to the above equations one would expect two different diazoamino-compounds to be formed thus—

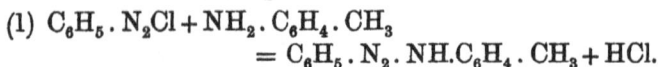

$$(1) \ C_6H_5 . N_2Cl + NH_2 . C_6H_4 . CH_3$$
$$= C_6H_5 . N_2 . NH.C_6H_4 . CH_3 + HCl.$$

---

* *Ber.*, 1894, **27**, 708.
† Bernthsen and Goske, *Ber.*, 1887, **20**, 925; Bamberger and Wulz, *Ber.*, 1891, **24**, 2082.

(2) $CH_3 . C_6H_4 . N_2Cl + NH_2 . C_6H_5$
$$= CH_3 . C_6H_4 . N_2 . NH.C_6H_5 + HCl.$$

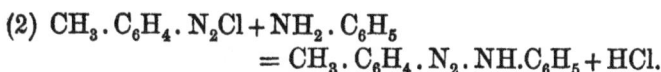

If one supposes, however, that an intermediate product is formed of the formula

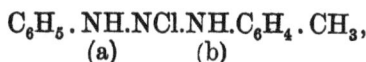

$$\underset{(a)}{C_6H_5 . NH.NCl.NH.}\underset{(b)}{C_6H_4 . CH_3},$$

then, by the elimination of hydrogen chloride, either of the above formulae is obtained according as to which hydrogen atom (a) or (b) is removed.*

In order to decide which of the above formulae is correct, use is made of the compound with phenylcarbimide. This combines with the diazoamino-compound to form a substance of formula—

(1) $C_6H_5 . NH.CO.N{\displaystyle <}^{C_6H_4 . CH_3}_{N_2 . C_6H_5}$

or (2) $C_6H_5 . NH.CO.N{\displaystyle <}^{C_6H_5}_{N_2 . C_6H_4 . CH_3}.$

When this is decomposed with dilute sulphuric acid, phenyl-p-tolylcarbamide, phenol, and nitrogen are formed, so that its constitution must be represented by (1), for (2) would give diphenylcarbamide. This conclusion is also confirmed by the fact that whichever way the compound is prepared it yields only one acetyl derivative, namely, diazobenzene p-aceto-toluidide, which, when decomposed by acids, yields aceto-toluidide.† The constitution of the diazoamino-compound is therefore $C_6H_5 . N_2 . NH.C_6H_4 . CH_3$,‡ and it is found that in these reactions the imino-group is always attached to the electronegative, and the diazo-group to the electropositive nucleus.

When the alkyl derivatives of the mixed diazoamino-compounds (that is, compounds in which the two radicals combined with the group N_3H are different) are examined, it is found that three isomeric substances exist. These are formed:—

I. By the action of $X.N_2Cl$ on $Y.NH.R$.

II. By the action of $Y.N_2Cl$ on $X.NHR$.

* V. Meyer, *Ber.*, 1881, **14**, 2447; 1888, **21**, 1016, 3004.
† von Pechmann, *Ber.*, 1895, **28**, 869.
‡ Goldschmidt, *Ber.*, 1888, **21**, 2578.

III. By the alkylation of X.N₃HY with RI and caustic potash. X and Y represent the two radicals united with the group N_3H, and R represents a univalent alkyl-group.

The isomerides obtained by direct alkylation are also formed when the compounds obtained according to (I) and (II) are heated together in equimolecular proportions.*

Migration of the diazo-group.—An interesting variation in this reaction is that in which the migration of the diazo-group occurs. Thus when diazotized sulphanilic acid and p-toluidine hydrochloride are mixed together at 0° the diazo- and amino-groups change places, and there results a mixture of p-diazo-toluene chloride and sulphanilic acid.† In neutral solution, however, the normal diazoamino-compound,

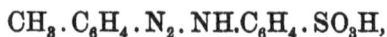

$$CH_3.C_6H_4.N_2.NH.C_6H_4.SO_3H,$$

is formed.

A corresponding interchange takes place between m- or p-nitrodiazobenzene chloride and p-toluidine. When, however, p-diazotoluene chloride is mixed with m- or p-nitro-aniline or sulphanilic acid no migration of the diazo-group takes place.‡ If diazobenzene chloride and p-bromoaniline are allowed to interact, aniline and p-bromodiazobenzene chloride are formed.§

This migration of the diazo-nucleus is probably associated with the changes which occur when this group passes from the 'diazonium' (see chap. xviii) to the diazo condition.

Bamberger ‖ found that when an alkali iso-diazo-oxide is dissolved in cold mineral acid, nitrous acid is formed—

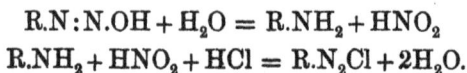

$$R.N:N.OH + H_2O = R.NH_2 + HNO_2$$
$$R.NH_2 + HNO_2 + HCl = R.N_2Cl + 2H_2O.$$

It is interesting to note that diazoamino-compounds may be obtained without the use of diazo-compounds; thus they

---

* Meldola and Streatfeild, *Trans.*, 1886, **49**, 624 ; 1887, **51**, 102, 434 ; 1888, **53**, 664 ; 1889, **55**, 412 ; 1890, **57**, 785.

† Griess, *Ber.*, 1882, **15**, 2190.

‡ Schraube and Fritsch, *Ber.*, 1896, **29**, 287.

§ Hantzsch and F. M. Perkin, *Ber.*, 1897, **30**, 1412.

‖ *Ber.*, 1895, **28**, 826.

are formed by the interaction of nitrosoamines and primary aromatic amines; for example—

$$C_6H_5(CH_3)N.NO + H_2N.C_6H_5 = C_6H_5(CH_3)N.N_2.C_6H_5.$$

Nitrosoacetanilide also reacts in a similar way—

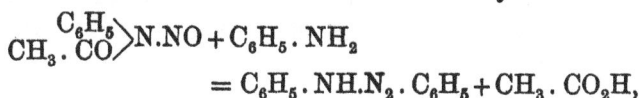

$$CH_3.\underset{CO}{\overset{C_6H_5}{>}}N.NO + C_6H_5.NH_2$$
$$= C_6H_5.NH.N_2.C_6H_5 + CH_3.CO_2H,$$

and two molecules react with one molecule of aniline in alkaline solution to form a bisdiazoamino-compound,

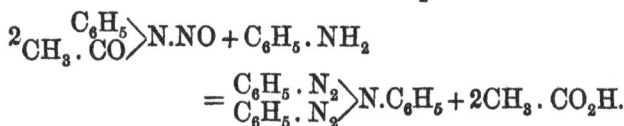

$$2CH_3.\underset{CO}{\overset{C_6H_5}{>}}N.NO + C_6H_5.NH_2$$
$$= \underset{C_6H_5.N_2}{\overset{C_6H_5.N_2}{>}}N.C_6H_5 + 2CH_3.CO_2H.$$

An isomeride of diazoaminobenzene is said to result when aniline is diazotized in presence of acetic acid instead of a mineral acid.* Its constitution is supposed to be

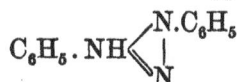

$$C_6H_5.NH\underset{N}{\overset{N.C_6H_5}{<}}$$

but the existence of such a compound must be accepted with reserve.†

Reactions of the diazoamino-compounds.—The diazoamino-compounds usually have a yellow colour, and do not dissolve in acids. They may generally be crystallized without decomposition, and are much more stable than the diazo-compounds.

When boiled with hydrochloric acid, nitrogen is evolved—

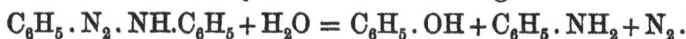

$$C_6H_5.N_2.NH.C_6H_5 + H_2O = C_6H_5.OH + C_6H_5.NH_2 + N_2.$$

On heating with cuprous chloride and hydrochloric acid, chlorobenzene and aniline are formed—

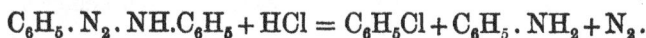

$$C_6H_5.N_2.NH.C_6H_5 + HCl = C_6H_5Cl + C_6H_5.NH_2 + N_2.$$

Hydrazines are obtained by reduction with zinc dust and acetic acid—

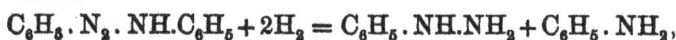

$$C_6H_5.N_2.NH.C_6H_5 + 2H_2 = C_6H_5.NH.NH_2 + C_6H_5.NH_2,$$

---

* Orloff, *J. Russ. Phys. Chem. Soc.*, 1906, **38**, 587.
† Compare also Vaubel, *Zeitsch. angew. Chem.*, 1900, **13**, 762; 1902, **15**, 1209.

and with nitrous acid two molecules of a diazo-compound are produced—

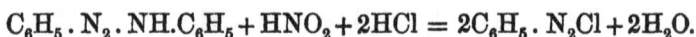

$$C_6H_5 . N_2 . NH.C_6H_5 + HNO_2 + 2HCl = 2C_6H_5 . N_2Cl + 2H_2O.$$

By boiling a diazoamino-compound with sulphurous acid in alcoholic solution, the diazo-group is replaced by the sulphonic acid group—

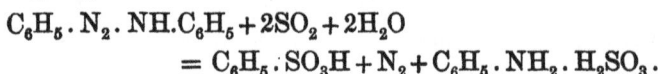

$$C_6H_5 . N_2 . NH.C_6H_5 + 2SO_2 + 2H_2O$$
$$= C_6H_5 ; SO_3H + N_2 + C_6H_5 . NH_2 . H_2SO_3 .$$

A very important reaction is that which takes place when a diazoamino-compound is warmed with a mixture of an amine and its hydrochloride; a molecular change occurs with formation of aminoazo-compounds—

$$C_6H_5 . N_2 . NH.C_6H_5 \;\rightarrow\; C_6H_5 . N_2 . C_6H_4 . NH_2 .$$

The velocity of the transformation of diazoamino- into aminoazo-compounds under the influence of aniline hydro-chloride has been shown by Goldschmidt and his pupils to be in accordance with the law of unimolecular reactions, in which

$$k = \frac{1}{t}\log - \frac{a}{a-x}. \quad *$$

In the case of diazoaminobenzene dissolved in aniline con-taining aniline hydrochloride, the rate of reaction is pro-portional to the concentration of the aniline hydrochloride, and increases with the temperature; as is usual in the case of a unimolecular reaction, the velocity is independent of the concentration of the diazoaminobenzene.

The transformation is also effected by other aniline salts, such as the dichloroacetate or trichloroacetate, but the rate in this case is slower than when the hydrochloride is used.†

Benzenediazoamino-p-toluene becomes converted into diazo-aminobenzene and p-toluidine, the former of which then undergoes transformation in accordance with the above rules.

---

* *Ber.*, 1896, **29**, 1369, 1899; *Zeitsch. physikal. Chem.*, 1899, **29**, 89.
† Compare also Jungius, *Chem. Weekblad*, 1905, **2**, 246.

The conversion takes place more slowly when the diazo-group is in the ortho-position with respect to the amino-group. Thus the value of the constant in the case of diazo-aminobenzene at 45° is 0·081, whilst the corresponding value for diazoamino-p-toluene is only 0·0095, the solution in each case being semi-normal.

CHAPTER XIII

AZO-COMPOUNDS

THE azo-compounds, like the diazo-, contain the group .N_2., but with the important difference that, whereas in the latter only one organic radical is united to the .N_2. group, the other free linking being combined with an acid radical, thus $R.N_2.Cl$; in the former two organic radicals are united to the N_2 group, thus $R.N_2.R$.

The groups attached to the nitrogen atoms may be either (1) aromatic, (2) aliphatic, or (3) one aromatic and one aliphatic group, giving the mixed azo-compounds.

The first representative of this class of compounds was obtained by Mitscherlich * by the distillation of nitrobenzene with alcoholic potash. Mitscherlich called the substance 'azobenzide', the modern name being of course azobenzene, $C_6H_5.N_2.C_6H_5$. He considered that the substance was formed by the replacement of one atom of nitrogen for one atom of hydrogen in benzene, but Zinin showed that 'azoxybenzide' was always formed in this reaction, and this on distillation yielded 'azobenzide'.† Zinin also, by the reduction of this substance with hydrogen sulphide and treatment of the product with sulphuric acid, obtained the sulphate of benzidine, which he considered was formed by the direct reduction of 'azobenzide'. Hofmann, however,‡ showed that in this reduction, hydrazobenzene, $C_6H_5.NH.NH.C_6H_5$, was first formed, which, in presence of sulphuric acid, underwent molecular change into the isomeric benzidine, $NH_2.C_6H_4.C_6H_4.NH_2$, and also proved by a vapour density determination that 'azobenzide' must have a formula double that which had been previously given to it.

* *Annalen*, 1834, 12, 311. † *J. pr. Chem.*, 1845, 36, 93.
‡ *Jahresber.*, 1863, 424.

The azo-compounds occupy an intermediate position between the nitro-compounds and the corresponding amines; thus in the case of nitrobenzene we have—

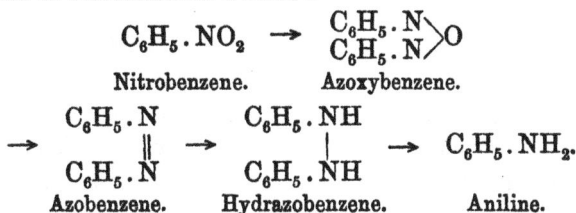

$$C_6H_5 . NO_2 \rightarrow \begin{matrix} C_6H_5 . N \\ C_6H_5 . N \end{matrix} \!\!\! > O$$

Nitrobenzene.　　　Azoxybenzene.

$$\rightarrow \begin{matrix} C_6H_5 . N \\ \| \\ C_6H_5 . N \end{matrix} \rightarrow \begin{matrix} C_6H_5 . NH \\ | \\ C_6H_5 . NH \end{matrix} \rightarrow C_6H_5 . NH_2.$$

Azobenzene.　　Hydrazobenzene.　　　Aniline.

§ 1. **Azoxy-compounds.**—These are obtained by the reduction of nitro- or nitroso-compounds with methyl- or ethyl-alcoholic potash—

$$4C_6H_5 . NO_2 + 3CH_3 . ONa = 2(C_6H_5 . N)_2O + 3H . CO_2Na + 3H_2O.$$

Other reducing agents which may be employed are sodium amalgam and alcohol, zinc dust and alcoholic ammonia, and arsenious acid in alkaline solution.

Further, azoxy-compounds are obtained by oxidizing amino- and azo-compounds with alkaline potassium permanganate or ferricyanide and by the oxidation of β-phenylhydroxylamine, $C_6H_5 . NH.OH$, in the air.

By treatment with energetic reducing agents, azoxy-compounds yield products of various degrees of reduction; thus with iron filings azo-compounds are formed, with ammonium sulphide, hydrazo-compounds, and, finally, acid reducing agents furnish amino-compounds; the acid used in the latter case obviously serves to bring about a molecular change in the hydrazo-compound first formed.

The simplest member of the series, azoxybenzene, is prepared as follows. 10 parts of sodium are dissolved in 100 parts of methyl alcohol, 15 parts of nitrobenzene added, and the whole heated for 3 hours on a boiling-water bath in an apparatus connected with an inverted condenser. The alcohol is then distilled off, the residue of sodium formate and azoxybenzene extracted with water, and the azoxybenzene allowed to crystallize out; the yield is about 90 to 92 per cent. of the theoretical.* Azoxybenzene is insoluble in water, but crystallizes from alcohol in long, yellow, rhombic needles melting at

* *Ber.*, 1882, 15, 865 ; 1883, 16, 81.

36°. On heating with concentrated sulphuric acid it undergoes isomeric change and is converted into hydroxyazobenzene—

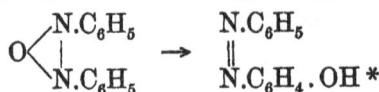

$$O\left\langle\begin{array}{l}N.C_6H_5 \\ | \\ N.C_6H_5\end{array}\right. \rightarrow \begin{array}{l}N.C_6H_5 \\ \| \\ N.C_6H_4.OH *\end{array}$$

§ 2. Azo-compounds.—As already indicated these are obtained by the reduction of nitro- or azoxy-compounds. The reducing agents used are, in the case of nitro-compounds, (1) zinc dust and alcoholic potash; (2) sodium amalgam and alcohol; (3) stannous chloride dissolved in caustic soda, and, in the case of azoxy-compounds, iron filings (p. 81).

Other methods of formation are: (1) by the oxidation of hydrazo- or amino-compounds by potassium permanganate in alkaline solution † or by potassium ferricyanide; (2) by the interaction of nitrosobenzene and aniline,

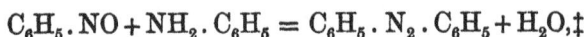

$$C_6H_5.NO + NH_2.C_6H_5 = C_6H_5.N_2.C_6H_5 + H_2O,\ddagger$$

of nitrosobenzene and phenylhydrazine,§ or of nitrosobenzene and s-diphenylhydrazine.‖ Azobenzene is also formed when phenylhydrazine is treated with bleaching powder solution,¶ or when diazo-compounds are treated with the same reagent; thus diazotized sulphanilic acid yields 2 : 2′-dinitroazobenzene-4 : 4′-disulphonic acid, $SO_3H.C_6H_3(NO_2).N : N.C_6H_3(NO_2).SO_3H$; some 4 : 6-dichloro-2-nitroaniline is formed at the same time, and this substance is also produced when diazobenzene chloride is treated with bleaching powder.

When the sodium *iso*-diazo-oxide derived from diazosulphanilic acid is used, the calcium salt of *p*-nitroaminobenzenesulphonic acid, $NO_2.NH.C_6H_4.SO_3H$, is formed.**

An interesting synthetic method for the production of these substances is that due to Meldola ;†† trinitroacetylaminophenol

  * *Ber.*, 1880, **13**, 525 ; *Annalen*, 1882, **215**, 218.
  † *Annalen*, 1867, **142**, 364.
  ‡ Baeyer, *Ber.*, 1874, **7**, 1638 ; Bamberger, *Ber.*, 1893, **26**, 473, 483.
  § Mills, *Trans.*, 1895, **67**, 925.
  ‖ Bamberger, *Ber.*, 1900, **33**, 3508.
  ¶ *Ber.*, 1897, **30**, 284.
  ** Zincke, *Ber.*, 1895, **28**, 2948 ; Zincke and Kuchenbecker, *Annalen* 1903, **330**, 1 ; see also Lenz, *Annalen*, 1903, **330**, 370.
  †† *Trans.*, 1906, **89**, 1943.

is condensed with phenylhydrazine, forming a hydrazo-compound which passes by oxidation into the azo-compound thus—

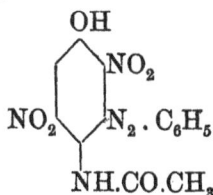

$$
\begin{array}{c}
\text{OH} \\
\text{NO}_2 \quad \overset{\displaystyle \nearrow}{\underset{\displaystyle\bigvee}{\bigcirc}} \text{NO}_2 \\
\text{N}_2 \,.\, \text{C}_6\text{H}_5 \\
\text{NH.CO.CH}_3
\end{array}
$$

The azo-compounds are usually strongly coloured owing to the presence of the chromophoric group .N':N.* They are readily acted on by sulphuric or nitric acids, chlorine, &c.

Azobenzene is prepared from azoxybenzene by distilling a mixture of one part of the latter with three parts of iron filings from a small retort. Care must be taken that the materials are quite dry. The reddish distillate is crystallized from light petroleum and forms red plates melting at 68° and boiling at 293°.

Complex azo-compounds are also obtained by treating diazo-salts with potassium ferrocyanide.†

§ 3. **Aminoazo-compounds.**—These are formed by the following reactions ‡ :—

(1) Intramolecular rearrangement of diazoamines—

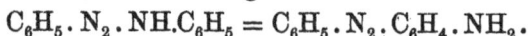

$$C_6H_5 . N_2 . NH.C_6H_5 = C_6H_5 . N_2 . C_6H_4 . NH_2 .$$

(2) Reduction of p-nitroazo-derivatives by an alkaline sulphide.§

(3) Combination of a diazotized monoacetyldiamine with an amine or phenol and subsequent hydrolysis of the acetyl-derivative.‖

(4) The action of nitrosobenzene on monoacetyldiamines and hydrolysis of the acetyl-derivative.¶

(5) The alkaline reduction of nitroamines.**

* See, however, Baly and Tuck, *Trans.*, 1906, **89**, 982.
† Griess, *Annalen*, 1866, **137**, 39 ; *Ber.*, 1876, **9**, 132 ; Ehrenpreis, *Bull. Acad. Sci. Cracow*, 1906, 265.
‡ Meldola and Eynon, *Trans.*, 1905, **87**, 1.
§ Meldola, *Trans.*, 1883, **43**, 425.
‖ Nietzki, *Ber.*, 1884, **17**, 343.   ¶ Mills, *Trans.*, 1895, **67**, 928.
** Haarhaus, *Annalen*, 1865, **135**, 164 ; Mixter, *Amer. Chem. J.*, 1883, **5**, 283 ; Nietzki, *Ber.*, 1884, **17**, 345 ; Gräff, *Annalen*, 1885, **229**, 341 ; Noelting and Binder, *Ber.*, 1887, **20**, 3016 ; Meldola and Andrews, *Trans.*, 1896, **69**, 10 ; Noelting and Fourneaux, *Ber.*, 1897, **30**, 2938.

(6) Combination of a monodiazo-chromate, prepared from a diamine by diazotizing only one amino-group and precipitating with sodium dichromate, with an amine or phenol.†

(7) Interaction of certain amines (the naphthylamines and 2 : 4-tolylenediamine, for example) and the para-diazoimides.‡

The usual method of preparation, however, is to allow a diazo-salt to react with amine in presence of sodium acetate; thus diazobenzene chloride unites with a solution of α-naphthylamine hydrochloride to form benzeneazo-α-naphthylamine—

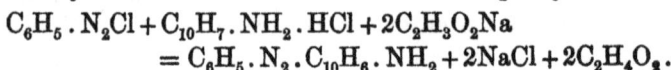

$$C_6H_5 . N_2Cl + C_{10}H_7 . NH_2 . HCl + 2C_2H_3O_2Na$$
$$= C_6H_5 . N_2 . C_{10}H_6 . NH_2 + 2NaCl + 2C_2H_4O_2 .$$

It is a fairly general rule that the diazo-group enters the aminic nucleus in the para-position with respect to the amino-group when this is unoccupied, but otherwise it enters the ortho-position.

In the following formulae the position of the entering diazo-group is shown by the asterisk.

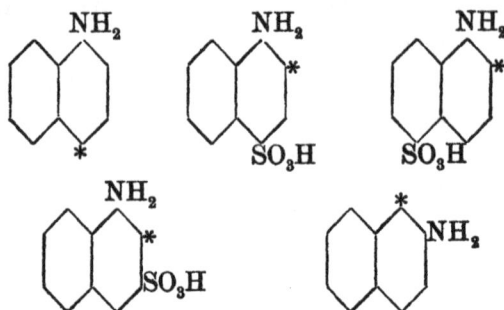

In the case of diamines combination only takes place with the meta- derivative; thus m-phenylenediamine and m-tolylenediamine

and

form azo-dyes, the diazo-group entering the para-position

† Meldola and Eynon, loc. cit.
‡ Morgan and Micklethwait, *Trans.*, 1907, **91**, 1512.

with respect to an amino-group. It is even possible for a second diazo-complex to combine at the carbon atom between the amino-groups. The question of the influence of substitution in such diamines on the formation of aminoazo-compounds has been made the subject of comprehensive researches by Morgan, who has arrived at the following conclusions * :—

(1) The mono-substituted meta-diamines and the di-substituted meta-diamines, containing one free para-position with respect to an amino-group, react readily with diazo-salts to furnish para-aminoazo-dyestuffs,† and this reaction takes place with equal readiness both with the primary meta-diamines of this type and with their completely alkylated derivatives.‡

(2) The di-para-substituted primary meta-diamines

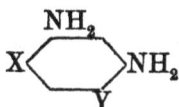

react with diazo-salts to form ortho-aminoazo-derivatives, but the reaction takes place much less readily than with those diamines having one free para-position, and the yield of azo-product is frequently very small.§

(3) The nature of the substituents X and Y exerts some influence on the course of the azo-condensation, for when they are methyl-groups the base (4 : 6-diamino-m-xylene) reacts with diazotized aniline and its homologues, but when both substituents are halogen atoms (chlorine, bromine, or iodine) the condensation does not occur with these simple diazo-salts, but only with those derived from the nitroanilines. When only one methyl-group is replaced by chlorine or bromine, reaction with diazotized aniline and p-toluidine still occurs, but the yield of o-aminoazo-derivative is extremely small.‖

(4) The presence of a nitro-group in one of the two substituted para-positions facilitates the azo-condensation, particularly when the diazo-salt also contains a substituent nitro-group.¶

* Trans., 1907, 91, 370. † Trans., 1900, 77, 1205 ; 1902, 81, 89.
‡ Trans., 1902, 81, 656. § Trans., 1902, 81, 1379 ; 1905, 87, 935.
‖ Trans., 1902, 81, 1379 ; 1905, 87, 937. ¶ Trans., 1905, 87, 940.

(5) The progressive alkylation of the di-para-substituted meta-diamines rapidly reduces their capacity for forming azo-derivatives. The symmetrically and unsymmetrically dimethylated diamines give mixtures of diazoamines and aminoazo-compounds,*, whilst the trimethylated diamines readily furnish diazoamines and show scarcely any tendency to form azo-derivatives. Finally, the interaction of the di-para-substituted m-diamines and diazo-salts is entirely prevented by the complete alkylation of these bases.†

The aminoazo-compounds are usually crystalline and soluble in alcohol. They are yellow to red or brown in colour, and many of them are used as dyestuffs.

The simplest member, aminoazobenzene, was introduced into commerce in 1863 by Simpson, Maule, and Nicholson under the name of 'aniline yellow'; at the present day it is only used as an intermediate product for the manufacture of induline, &c., but its sulphonic acids find extensive application as 'fast yellow'.

§ 4. **Hydroxyazo-compounds.**—These are obtained by the action of a diazo-salt on phenols and their derivatives in alkaline solution,‡ thus—

$$C_6H_5 . N_2Cl + C_6H_5 . OH = C_6H_5 . N_2 . C_6H_4 . OH + HCl.$$

The same rules apply here with regard to the position taken by the diazo-complex as in the case of the aminoazo-compounds, namely, that the para-position to the hydroxyl-group, if free, is occupied. Otherwise the ortho-position is taken. In the case of β-naphthol, the diazo-group enters the α-position.

It has been found, however, that by the action of diazobenzene or p-diazotoluene on phenol, small amounts of the corresponding o-hydroxyazo-compounds are formed.§

* *Trans.*, 1905, **87**, 946; 1906, **89**, 1057; 1907, **91**, 368.
† *Trans.*, 1902, **81**, 656.
‡ In acid solution, diphenyl ether is produced,
$$C_6H_5 . N_2 . SO_4H + C_6H_5 . OH = (C_6H_5)_2O + H_2SO_4 + N_2$$
(Hofmeister, *Annalen*, 1871, **159**, 191), to the extent of a 4 per cent. yield.
§ Bamberger, *Ber.*, 1900, **33**, 3188.

With two molecules of a diazo-compound, the bisazo-derivative is formed

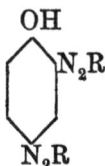

OH

N_2R

N_2R

and with three molecules the trisazo-compound

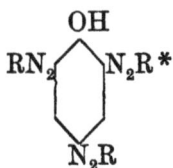

OH

RN_2 N_2R *

N_2R

α-Naphthol yields with one molecule of a diazo-salt, first, the monoazo-compound

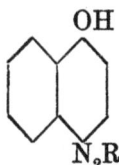

OH

N_2R

and with two molecules the bisazo-compound

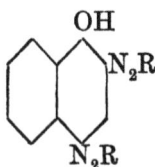

OH

N_2R

N_2R

Resorcinol, like phenol and *m*-phenylenediamine, can also combine with two molecules of a diazo-compound; thus with one molecule it yields

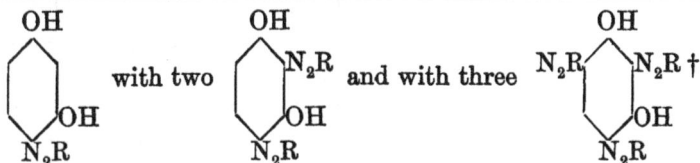

OH   OH   OH

with two N_2R and with three N_2R N_2R †

OH   OH   OH

N_2R N_2R N_2R

---

* Grandmougin and Freimann, *Ber.*, 1907, **40**, 2662; Heller and Nötzel, *J. pr. Chem.*, 1907 [ii], **76**, 58.
† Orndoff and Ray, *Ber.*, 1907, **40**, 3211.

For a more detailed description of the amino- and hydroxy-azo-dyestuffs a larger work must be consulted.*

§ 5. Rate of formation of amino- and hydroxyazo-compounds.—This has been measured by Goldschmidt and his pupils. The case of the formation of methyl-orange from *p*-diazobenzenesulphonic acid and dimethylaniline hydrochloride was studied first, the method of procedure being to withdraw samples of the mixture at given periods of time, and after acidifying, to determine the quantity of diazo-compound present by measuring the nitrogen evolved on boiling.

The conclusions arrived at were : (1) In the combination of the hydrochloride of a tertiary amine with diazobenzenesulphonic acid, it is the base liberated from the hydrochloride by hydrolysis that acts with the diazo-compound. (2) Excess of hydrochloric acid retards the combination. (3) The concentration of the hydrochloride or of the diazo-compound has no influence on the reaction. (4) The combination proceeds at the same rate when other acids of the same strength are used (hydrobromic, nitric, &c.), but much more quickly when weak acids such as acetic, the chloroacetic acids, formic, propionic, levulinic, or lactic acids are employed. Investigations were also made as to the influence of the base used; comparisons of the velocity of formation of the azo-compound between dimethyl-, diethyl-, and dipropyl-aniline show that the replacement of methyl by ethyl lowers, and that of ethyl by propyl increases, the velocity. Dimethyl- and diethyl-*m*-toluidine combine more rapidly, and dimethyl- and diethyl-*m*-chloroanilines more slowly, than the corresponding unsubstituted compounds. In the formation of hydroxyazo-compounds from phenols and diazo-salts in alkaline solution the active agents are the free phenol liberated by the hydrolytic action of the water on the alkali salt and the *syn*-diazo-compound (for an explanation of this see p. 123). An excess of alkali retards the combination, and the more con-

* See Bülow, *Chemische Technologie der Azo-Farbstoffe*, 1897 ; Pauli, *Synthese der Azo-Farbstoffe*, 1904 ; Cain and Thorpe, *Synthetic Dyestuffs*, 1905, p. 47.

centrated the solution of the phenol and diazo-salt the more
time is required for the combination. Similar results were
obtained when sodium diazo-oxide (see p. 99) was used instead
of the diazo-salt.

§ **6. Constitution of the hydroxyazo-compounds.**—The earlier
assumptions that the hydroxyazo-compounds possessed the
formula R.N : N.R.OH were somewhat shaken by the
discovery that, by the action of phenylhydrazine on α- and
β-naphthaquinones, compounds resulted which were identical
with those obtained by treating α- and β-naphthol with diazo-
salts,* and it was found that phenylhydrazine and benzo-
quinoneoxime yielded a condensation product which was
identical with p-hydroxyazobenzene, thus—

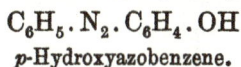

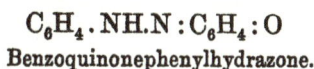

$$C_6H_5 . N_2 . C_6H_4 . OH \qquad C_6H_4 . NH.N : C_6H_4 : O$$
p-Hydroxyazobenzene.   Benzoquinonephenylhydrazone.

The question was attacked for many years chiefly from the
purely chemical side. Goldschmidt and his pupils,† McPher-
son,‡ and Hewitt,§ expressed the opinion that para-hydroxyazo-compounds were true phenols, and ortho-compounds
quinones, whilst Jacobson,‖ Meldola,¶ and Nietzki and
Kostanecki ** adhered to the view that both series of compounds possessed the phenolic constitution.

On the other hand, Farmer and Hantzsch,†† from determinations of electrical conductivity, and Möhlau and Kegel,‡‡ from
the reactions with carbinols, expressed the opinion that both
ortho- and para-hydroxyazo-compounds in the free state were
really quinones, the metallic salts, however, being phenolic in
character.

The views generally held about this time, therefore (1900–

* Zincke, *Ber.*, 1884, **17**, 3026 ; 1887, **20**, 3171 ; compare also Lieber-
mann, *Ber.*, 1883, **16**, 2858.
† *Ber.*, 1890, **23**, 487 ; 1891, **24**, 2300 ; 1892, **25**, 1324.
‡ *Ber.*, 1895, **28**, 2414 ; *Amer. Chem. J.*, 1899, **22**, 364 ; Auvers, *Ber.*,
1896, **29**, 2361 ; 1900, **33**, 1302.
§ *Trans.*, 1900, **77**, 99, 712.   ‖ *Ber.*, 1888, **21**, 414.
¶ *Trans.*, 1888, **53**, 460 ; 1889, **55**, 114, 603 ; 1891, **59**, 710 ; 1893, **63**,
923 ; 1894, **65**, 834.
** *Ber.*, 1890, **23**, 3263 ; 1891, **24**, 1592, 3977.
†† *Ber.*, 1899, **32**, 3089.   ‡‡ *Ber.*, 1900, **33**, 2858.

1903), were that the metallic salts and alkyl ethers of all hydroxyazo-compounds, as well as the acyl-derivatives of p-hydroxyazo-compounds, were true azo- (phenolic) compounds.

It was considered, further, that the free ortho-compounds were quinonoid in constitution, as were also the acid additive products of both series of compounds, for example—

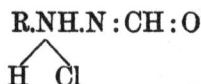

$$R.NH.N:CH:O$$
$$\overset{\wedge}{H\ \ Cl}$$

Opinions as to the constitution of the free para-hydroxy-compounds were, however, divided, as has been shown.

From this period onwards a number of researches rather tend to show that the phenolic constitution for both the free ortho- and para-compounds is to be adopted.

Thus the properties of m-hydroxyazophenol were found by Jacobson and Hönigsberger to agree closely with those of the para-derivative,* and these chemists concluded that the free para-compounds, as well as their additive compounds with acids, were to be regarded as azo (phenolic) in constitution.

This view was confirmed by the researches of Borsche,† and Goldschmidt and Löw-Beer showed that the earlier opinion of Goldschmidt as to the quinonoid character of the ortho-compounds had been based on incorrect data, and concluded that all hydroxyazo-compounds possessed the azo (phenolic) structure.

Most of the later work has confirmed this conclusion,‡ although Tuck, from an examination of the absorption spectra, inclines to the view that the free ortho-compounds are quinonoid, whilst the para-compounds, their hydrochlorides, and the hydrochlorides of the ortho-compounds, are phenolic in structure.§

---

* *Ber.*, 1903, **36**, 4093.
† *Annalen*, 1904, **334**, 143; 1905, **340**, 85; 1907, **357**, 171.
‡ Hewitt and Mitchell, *Proc.*, 1905, **21**, 298; Mitchell, *Trans.*, 1905, **87**, 1229; Willstätter and Veraguth, *Ber.*, 1907, **40**, 1432; Auvers, *Ber.*, 1907, **40**, 2154.
§ *Trans.*, 1907, **91**, 449; compare also Hewitt and Mitchell, *Trans.*, 1907, **91**, 1251.

§ **7. Mixed azo-compounds.**—These are represented by the general formula A.N : N.B where A is an aliphatic and B an aromatic group.

The simplest member of this class, namely, benzeneazomethane, is best prepared by treating phenylhydrazine with formaldehyde.* It is a yellow, volatile oil, boiling at 150° with decomposition.

The first representative of this class, however, prepared directly from a diazo-compound was obtained by V. Meyer and Ambühl; † by the action of diazobenzene nitrate on sodium nitroethane they prepared benzeneazonitroethane—

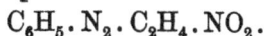

$$C_6H_5.N_2.C_2H_4.NO_2.$$

This is probably not a true azo-compound, but a hydrazone-derivative having the constitution—

$$C_6H_5.N_2H : C_2H_3.NO_2.$$

Shortly afterwards Friese ‡ described benzeneazonitromethane as being obtained by treating sodium nitromethane with diazobenzene nitrate, but Bamberger has shown that Friese's compound was nitroformazyl, produced according to the equation

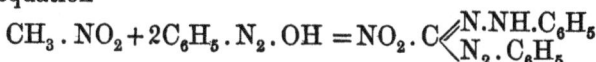

$$CH_3.NO_2 + 2C_6H_5.N_2.OH = NO_2.C{\underset{N_2.C_6H_5}{\overset{N.NH.C_6H_5}{<}}}$$

This behaviour of nitromethane is exceptional, as formazyl-derivatives are not produced with the higher homologues of nitromethane; under certain conditions, however, the simple compound nitroformaldehydrazone, $NO_2.CH : N_2H.C_6H_5$, is obtained.§

By the action of diazo-salts on the sodium-derivative of acetoacetic ester V. Meyer obtained a compound which was first considered to be a true mixed azo-compound, thus—

$$CH_3.CO.CH(CO_2Et).N_2.C_6H_5,\|$$

but later he considered that, owing to its supposed insolubility in alkali, the substance possessed a hydrazone structure—

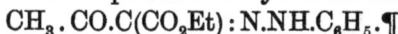

$$CH_3.CO.C(CO_2Et) : N.NH.C_6H_5.¶$$

---

* Baly and Tuck,*Trans.*,1906,**89**,986; see also Tafel,*Ber.*,1885,**18**,1742.
† *Ber.*, 1875, **8**, 751, 1053.   ‡ *Ber.*, 1875, **8**, 1078.
§ *Ber.*,1894, **27**, 155 ; 1900, **33**, 2043 ; compare also Oddo and Ampola, *Gazzetta*, 1893, **23**, i. 257, β naphthylazonitroethane.
‖ *Ber.*, 1876, **9**, 384 ; 1878, **11**, 1418 ; 1884, **17**, 1928.
¶ *Ber.*, 1888, **21**, 12.

He adhered to this opinion, although he found that the compound was really soluble in alkali.* Kjellin also adopted this view, and noticed that apparently two isomeric condensation products were formed; these he considered to be stereoisomeric hydrazones, thus—

$$CH_3 . CO.C.CO_2Et \qquad\qquad CH_3 . CO.C.CO_2Et$$
$$C_6H_5 . NH.N \qquad\qquad\qquad N.NH.C_6H_5 \dagger$$

Bülow, however, showed that not only was the compound very readily soluble in alkali, but that it could not be acetylated and was not acted on by benzoyl chloride or methyl iodide, so that the original theory of V. Meyer was correct, and the compound must be regarded as the true azo-derivative

$$CH_3 . CO.CH(CO_2Et).N : N.C_6H_5 . \ddagger$$

Bülow obtained the same compound from diazobenzene chloride and sodium benzene-*iso*-diazo-oxide.§

By the action of a diazo-salt on the sodium-derivative of ethyl methylacetoacetate, the acetyl-group is eliminated and the compound was written $CH_3 . CH(CO_2Et).N : N.C_6H_5$,‖ the corresponding acid being $CH_3 . CH(CO_2H).N_2 . C_6H_5$. This, however, was found to be identical with the condensation product formed by the action of phenylhydrazine on pyroracemic acid, $CH_3 . C(CO_2H) : N.NH.C_6H_5$.¶

In order to decide which of these formulae was correct Japp and Klingemann ** treated the so-called benzeneazoacetone †† with sodium ethoxide and ethyl chloroacetate, and reduced the corresponding acid. The substance obtained was anilinoacetic acid, proving that the . CH_2 . CO_2H group had combined with the nitrogen atom attached to the phenylgroup. The second formula for benzeneazoacetone is there-

* *Ber.*, 1888, **21**, 2121.
† *Ber.*, 1897, **30**, 1965; compare also Favrel, *Compt. rend.*, 1901, **132**; 983; 1898, **127**, 116.
‡ *Ber.*, 1899, **32**, 197.
§ *Ber.*, 1898, **31**, 3122.
‖ Japp and Klingemann, *Ber.*, 1887, **20**, 2942, 3284, 3398.
¶ E. Fischer and Jourdan, *Ber.*, 1883, **16**, 2241.
** *Trans.*, 1888, **53**, 521, 538.
†† V. Meyer and Münzer, *Ber.*, 1878, **11**, 1417,
$\qquad CH_3 . CO.CH_2 . N_2 . C_6H_5$ or $CH_3 . CO.CH : N.NH.C_6H_5$.

fore the correct one, and the compound is really a mono-
hydrazone of pyruvic aldehyde. This was also confirmed by
the fact that it yields an osazone with phenylhydrazine.
This conclusion is, however, not to be applied universally,
for when diazobenzene chloride acts on the sodium derivative
of acetoacetaldehyde, benzeneazoacetaldehyde,

$$CH_3 . CO.CH(CHO).N_2 . C_6H_5,$$

is formed, and this, with phenylhydrazine, yields the corresponding hydrazone—

$$CH_3 . CO.CH(CH:N.NH.C_6H_5).N_2 . C_6H_5.†$$

When diazobenzene chloride acts on acetonedicarboxylic
acid in presence of sodium acetate, the bishydrazone of
mesoxalaldehyde, $CO(CH:N.NH.C_6H_5)_2$, is obtained.‡ If,
however, the ethyl ester of acetonedicarboxylic acid is subjected to the action of p-nitrodiazobenzene, a compound is
obtained which reacts partly as a hydrazone and partly as an
azo-derivative. This is therefore to be regarded as containing a labile hydrogen atom, indicated by a star in the
formula—

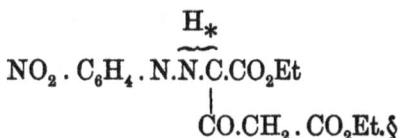

$$NO_2 . C_6H_4 . \overset{\displaystyle H_*}{\overbrace{N.N}}.C.CO_2Et$$
$$\underset{\displaystyle CO.CH_2 . CO_2Et.§}{\big|}$$

When a hydrazone such as is described above, in which the
group $C_6H_5 . NH.N:C:$ is combined with H, CO_2H, or COR,
each of the latter groups can be replaced by the action of
diazobenzene in alkaline or acetic acid solution. Thus, by
the action of diazobenzene on malonic ester, V. Meyer and
Münzer ‖ also obtained a condensation product which they
regarded as

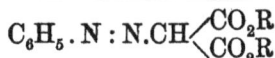

$$C_6H_5 . N : N.CH {\begin{matrix} \diagup CO_2R \\ \diagdown CO_2R \end{matrix}}$$

† Claisen and Beyer, *Ber.*, 1888, **21**, 1697; compare also *Ber.*, 1892,
25, 3190.
‡ v. Pechmann and Jenisch, *Ber.*, 1891, **24**, 3255; v. Pechmann and
Vanino, *Ber.*, 1892, **25**, 3190.
§ Bülow and Höpfner, *Ber.*, 1901, **34**, 71; compare also Bülow and
Hailer, *Ber.*, 1902, **35**, 915.
‖ Loc. cit.

but which is now considered to be the phenylhydrazone of malonic ester—

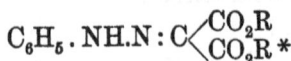

$$C_6H_5 . NH.N : C\begin{cases}CO_2R\\CO_2R\end{cases}*$$

If the acid itself is used, formazylcarboxylic acid and formazyl result †—

$$C_6H_5 . NH.N : C\begin{cases}N_2 . C_6H_5\\CO_2H\end{cases}$$ $$C_6H_5 . NH.N : C\begin{cases}N_2 . C_6H_5\\H\end{cases}$$

Formazylcarboxylic acid.   Formazyl.

The latter is also formed by the action of diazobenzene on the ethyl hydrogen salt of phenylhydrazonemalonic acid.

By the further action of diazobenzene on formazyl, or its carboxylic acid, formazylazobenzene

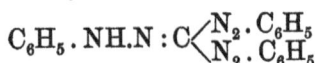

$$C_6H_5 . NH.N : C\begin{cases}N_2 . C_6H_5\\N_2 . C_6H_5\end{cases}$$

is obtained, as well as by allowing diazobenzene and acetaldehyde to interact in alkaline solution.‡

These compounds are usually dark red, crystalline substances, and on reduction they give colourless hydrazones.§

Formazylcarboxylic acid is also obtained by the action of diazobenzene on ethyl acetoacetate under certain conditions; the first product of the reaction being the hydrazone—

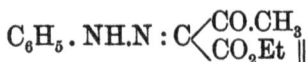

$$C_6H_5 . NH.N : C\begin{cases}CO.CH_3\\CO_2Et\end{cases}\|$$

By the action of diazobenzene on ethyl oxalacetate the hydrazone, $C_6H_5 . NH.N:C (CO_2Et).CO.CO_2Et$, is first formed, and with a further molecule of diazobenzene, ethyl diphenylformazylformate, $C_6H_5 . N:N.C(N.NH.C_6H_5).CO_2Et$, is produced.¶

Further confirmation of the hydrazone constitution of such condensation products is afforded by the action of diazo-

* Compare also Favrel, *Compt. rend.*, 1899, **128**, 829 ; 1901, **132**, 1336.
† Compare also Busch and Wolbring, *J. pr. Chem.*, 1905 [ii], **71**, 366.
‡ Bamberger and Müller, *Ber.*, 1894, **27**, 147.
§ v. Pechmann, *Ber.*, 1892, **25**, 3175.
‖ Bamberger, *Ber.*, 1892, **25**, 3201, 3539, 3547.
¶ Rabischong, *Bull. Soc. chim.*, 1904 [iii], **31**, 76, 83 ; compare also Favrel, *Compt. rend.*, 1899, **128**, 318.

benzene on ethyl cyanoacetate whereby the phenylhydrazone of the latter is produced.*

The question of the constitution of the mixed azo-compounds cannot, however, be said to be finally settled by fixed rules. Some appear to be tautomeric substances, others true hydrazones, and many undoubtedly true azo-compounds. In all probability the course of the reaction depends on the constitution of the diazo-compound used, the discussion of which is postponed until later (see p. 112).†

---

* Krückeberg, *J. pr. Chem.*, 1892 [ii], **46**, 579, **47**, 591, **49**, 321; compare also Uhlmann, ibid. **51**, 217 ; Marquardt, ibid. **52**, 160; Favrel, *Compt. rend.*, 1900, **131**, 190 ; 1907, **145**, 194.

† For mixed bisazo-compounds, see Duval, *Compt. rend.*, 1907, **144**, 1222.

CHAPTER XIV

METALLIC DIAZO-DERIVATIVES.
DIAZO-HYDROXIDES

BY adding a cold saturated solution of diazobenzene nitrate to a large excess of concentrated aqueous potassium hydroxide, evaporating the resulting yellow liquid, and extracting with alcohol, Griess obtained a substance containing potassium, to which he gave the formula $C_6H_5N_2 . OK$.

When this compound was treated with acetic acid a yellow oil was obtained which Griess regarded as free diazobenzene, $C_6H_4N_2$.

The potassium compound was examined by Curtius,* who found, however, that it contained only two-thirds of the nitrogen required by the above formula. Also when diazobenzene sulphate was neutralized with barium hydroxide and the mixture extracted with ether, a yellow substance was obtained which melted at $-3°$. This contained only two atoms of nitrogen to three benzene nuclei, although no nitrogen had been evolved as gas.

Griess's product was shown later by Bamberger to consist chiefly of the *iso*-salt.

A very important addition to the chemistry of the metallic diazo-derivatives was made by Schraube and Schmidt in 1894.† These chemists found that when a 10 per cent. solution of *p*-nitrodiazobenzene chloride was quickly added to an 18 per cent. solution of sodium hydroxide at 50–60°, golden-yellow plates separated which no longer combined with β-naphthol, and which they considered to be sodium *p*-nitrophenylnitrosoamine, $NO_2 . C_6H_4 . NNa.NO$.

When an ice-cold, aqueous solution of this was treated with

---

* *Ber.*, 1890, **23**, 3035.　　　　　† *Ber.*, 1894, **27**, 514.

acetic acid, a pale yellow precipitate was obtained which was regarded as the free p-nitrophenylnitrosoamine,

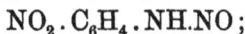

$$NO_2 . C_6H_4 . NH.NO;$$

this, like the sodium salt, did not couple with alkaline β-naphthol. When hydrochloric acid was substituted for acetic acid, p-nitrodiazobenzene chloride was slowly regenerated. On treatment with methyl iodide, the sodium salt gave the nitrosoamine of p-nitromonomethylaniline—

$$NO_2 . C_6H_4 . NMe.NO.$$

Schraube and Schmidt also investigated the properties of Griess's potassium salt, and showed that it differed from the compound described by them in that it coupled with alkaline β-naphthol. On being heated with concentrated aqueous potassium hydroxide, however, it was converted into the potassium salt of phenylnitrosoamine, $C_6H_5 . NK.NO$, which no longer combined with β-naphthol, and with methyl iodide gave the nitrosoamine of methylaniline. This potassium salt, on neutralization with acetic acid, gave a colourless oil which combined with β-naphthol solution, and on adding an excess of acetic acid to the oil, a solution of diazobenzene acetate was obtained.

From this work Schraube and Schmidt drew the following conclusions:—

(1) The alkali salts of diazobenzene can exist in two forms, namely, the diazo-metallic salts, $C_6H_5 . N : N.OR$, and the nitrosoamines, $C_6H_5 . NR.NO$ (R denoting the metal).

(2) Free phenylnitrosoamine, $C_6H_5 . NH.NO$, does not exist, as a solution of its non-combining sodium salt, when acidified with acetic acid, immediately gives an azo-dyestuff with β-naphthol.

(3) p-Nitrodiazobenzene appears to exist only in the nitrosoamine form, and its alkali salt does not exist in the 'oxime' condition.

These conclusions, as will be shown, do not truly represent the course of the reactions (see p. 98).

In a paper published shortly after the appearance of Schraube and Schmidt's work, Bamberger stated that he had earlier discovered a derivative of β-naphthylamine corresponding to the formula $C_{10}H_7.NH.NO$, which did not couple with β-naphthol, but did so after treatment with a mineral acid.*

He considered that the influence of the latter was to effect a transformation into the isomeric diazo-compound, thus—

$$C_{10}H_7.NH.NO \;\rightarrow\; C_{10}H_7.N:N.OH.$$

Confirmation of this view was afforded by von Pechmann and Frobenius, who stated that the silver salt of p-nitrophenylnitrosoamine, when treated with methyl iodide, yielded an oxygen-ether of p-nitrodiazobenzene—

$$NO_2.C_6H_4.N:N.O.CH_3.$$

The conclusion is emphasized, therefore, that the hydroxide corresponding with these compounds exhibits the phenomenon of tautomerism.† That is to say, that the hydroxide can act either as

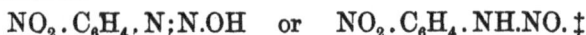

$$NO_2.C_6H_4.N:N.OH \quad\text{or}\quad NO_2.C_6H_4.NH.NO. \ddagger$$

It was thus established that two isomeric forms of the metallic diazo-compounds exist; the modification described by Schraube and Schmidt may be called the stable or *iso*-modification, and the labile or normal form is that which couples with phenols much more readily than its isomeride.§

Most of the metallic diazo-compounds exist in these two modifications, but the presence of negative groups in the aromatic nucleus greatly diminishes the stability of the normal modification. Although, for the purpose of defining these isomeric compounds, it has been necessary to mention the constitutions which were assigned to them at the time of their discovery, the subsequent developments of the views on this subject have been so extensive, and the discussion so prolonged, that an account of this must be postponed. We

* *Ber.*, 1894, **27**, 679.

† *Ber.*, 1894, **27**, 672; compare also Bamberger, *ibid.*, 679.

‡ Although Hantzsch would not accept the views of Bamberger and von Pechmann, yet he arrived at this conclusion from his own work some years later (p. 151).

§ The presence of alkali has a great effect on the combining power of the two isomerides: Schraube and Schmidt had a large excess of alkali present when they noticed that the stable form did not combine, but when less is used it does combine, although much more slowly than the labile form.

shall therefore proceed to a description of some of the more important compounds of this class.

Potassium benzenediazo-oxide (normal, labile, or *syn*-salt). —10 c.c. of a 15 per cent. solution of diazobenzene chloride are dropped slowly into a mixture of 140 grams of potassium hydroxide and 60 grams of water cooled to 5°. The temperature is allowed to rise to 15–20°, whereby the potash becomes completely dissolved, and the precipitated potassium benzenediazo-oxide is collected. This is pressed on porous porcelain. One gram of the crude product is now dissolved in 3 c.c. of absolute alcohol at −5°, the solution quickly filtered, and 8–10 times its volume of dry ether added. The salt is obtained in this way in snow-white, silky needles, which are very hygroscopic and soon become pink.*

syn-Diazo-oxides are also obtained by treating nitrosoacylanilides with potassium hydroxide,†

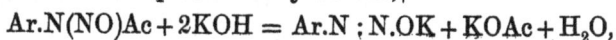

$$Ar.N(NO)Ac + 2KOH = Ar.N : N.OK + KOAc + H_2O,$$

and by the reduction of salts of diazoic acid,

$$Ar.N_2O.OK + 2H = H_2O + Ar.N : N.OK.$$

When the normal salts are treated with acids the corresponding hydroxides are not formed, but the yellow, explosive diazo-anhydrides are produced (see p. 148).

Potassium benzenediazo-oxide (*iso*-, stable, or *anti*-salt).— This is obtained by heating the strongly alkaline solution of the diazo-chloride to 130–140° until the product no longer combines with β-naphthol (compare p. 98).

Another method of preparing the *iso*-metallic compounds consists in treating *o*- or *p*-hydroxybenzylated nitrosoaryl-derivatives of the type $NO.NAr.CH_2.C_6H_4.OH$ with very dilute aqueous potassium hydroxide—

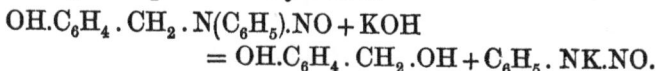

$$OH.C_6H_4.CH_2.N(C_6H_5).NO + KOH$$
$$= OH.C_6H_4.CH_2.OH + C_6H_5.NK.NO.$$

The potassium *iso*-diazo-oxide is formed together with hydroxybenzyl alcohol.‡

---

* Bamberger, *Ber.*, 1896, **29**, 461.
† Bamberger, *Ber.*, 1894, **27**, 915.
‡ Bamberger and Müller, *Annalen*, 1900, **313**, 97.

iso-Diazo-oxides are also formed by the reduction of *iso*-diazoic acids, and by heating secondary nitrosoamines with potassium hydroxide.*

When the potassium *iso*-diazo-oxides are treated with acetic or mineral acids, the corresponding hydroxides are obtained, which, with alkalis, regenerate the metallic *iso*-salt.

The hydroxides of *iso*-diazobenzene, *iso*-diazo-*p*-toluene, *p*-chloro- and *p*-bromo-*iso*-diazobenzene, α- and β-*iso*-diazonaphthalene, and potassium *iso*-diazobenzenesulphonate are all colourless, whilst the hydroxides of *o*- and *p*-nitro-*iso*-diazobenzene are yellow.†

iso-Diazobenzene hydroxide is a colourless oil which is readily soluble in ether. It is very unstable.

The hydroxides of the remaining substances mentioned above are white crystalline solids.‡

The action of alkalis on diazo-salts sometimes, however, proceeds differently. Thus 2:4:6-tribromodiazobenzene gives rise to 3:5-dibromo-*o*-benzoquinonediazide

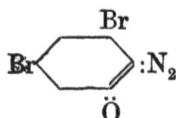

$$\text{Br} \diagdown \!\!\!\!\bigcirc\!\!\!\! \diagup \; :N_2$$
$$\overset{Br}{\underset{\ddot{O}}{}}$$

(see p. 67), and *o*-diazotoluene furnishes indazole—

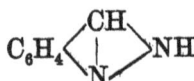

$$C_6H_4 \diagdown \!\!\!\begin{array}{c} CH \\ | \\ N \end{array}\!\!\! \diagup NH$$

(see p. 31).§

Diazobenzene hydroxide (diazonium hydroxide).—This hydroxide does not correspond with either of the two foregoing potassium salts according to Hantzsch, although Bamberger regards it as the hydroxide derived from the labile salt (see p. 144).

For the preparation, 0·7 gram of pure diazobenzene chloride is dissolved in about 50 c.c. of ice-cold water, and about

0·8 gram (the theoretical amount is 0·62) of freshly-precipitated moist silver oxide mixed with ice, added, and the whole shaken for five minutes. The filtrate consists of a practically pure solution of the hydroxide.

This solution has a strongly alkaline reaction, and combines instantly with β-naphthol. The pure solution is colourless. The hydroxide is also obtained by treating a solution of the . diazo-sulphate with the calculated amount of baryta.* The solution is very unstable, even at 0° it decomposes and becomes dark coloured.

Other diazo-hydroxides are prepared in a similar manner.

Reduction of the metallic diazo-oxides.—Both series of diazo-oxides, when treated with sodium amalgam in presence of excess of sodium hydroxide, yield the corresponding hydrazine.†

Oxidation of the metallic diazo-compounds. Aromatic diazoic-acids.—When an alkaline solution of diazobenzene is oxidized by potassium ferricyanide, a substance is obtained to which the name benzenediazoic acid is given—

$$C_6H_5 . N_2O_2H.$$

It is formed in white leaflets, melting at 46–46·5°, sparingly soluble in water, but readily so in organic solvents or alkalis. It forms well-defined salts.‡

Potassium permanganate may also be used as the oxidizing agent.§

A better yield is obtained by oxidizing potassium benzene-iso-diazo-oxide with potassium ferricyanide.‖ The compound may also be prepared by treating diazobenzene perbromide with sodium hydroxide,¶ or by the action of nitrogen pentoxide on aniline in ethereal solution at −20°.**

When benzenediazoic acid is slowly heated, or when it is

* Hantzsch, *Ber.*, 1898, **31**, 340.   † Hantzsch, *Ber.*, 1898, **31**, 340.
‡ Bamberger and Storch, *Ber.*, 1893, **26**, 471; Bamberger, ibid., 1894, **27**, 359.
§ Bamberger and Landsteiner, *Ber.*, 1893, **26**, 482.
‖ Bamberger, *Ber.*, 1894, **27**, 914.
¶ Bamberger, *Ber.*, 1894, **27**, 1273.
** Bamberger, *Ber.*, 1894, **27**, 584.

treated with mineral acids, it undergoes intramolecular change with formation of a mixture of o- and p-nitroaniline.

When heated with potassium hydroxide to about 290° it is decomposed into aniline and nitric or nitrous acid. By gentle reduction with zinc and acetic acid, diazobenzene is formed, and with sodium amalgam, phenylhydrazine is produced.*

On account of the conversion into nitroaniline, Bamberger regarded benzenediazoic acid as the simplest aromatic nitramine, or phenylnitramine, $C_6H_5 . NH.NO_2$, and represented the change into nitroaniline as follows—

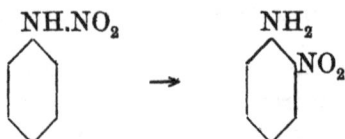

The proof of this constitution was found in the study of the action of hypochlorite on the diazoic acid, for a chloroderivative was obtained which gave the characteristic reactions of a chloroimide, and underwent molecular change even more readily than the parent compound, forming p-chloro-o-nitroaniline.

The constitution of the chloro-compound is therefore $C_6H_5 . NCl.NO_2$.†

Benzenediazoic acid forms two methyl esters; with methyl iodide the sodium salt gives the α-ester, $C_6H_5 . NMe.NO_2$, and the silver salt yields the β-ester, $C_6H_5 . N:NO_2Me$.

Benzenediazoic acid is therefore, as Bamberger had shown to be the case with diazobenzene hydroxide, a tautomeric substance, thus—

$$C_6H_5 . NH.NO_2 \rightleftarrows C_6H_5 . N:NO_2H.$$

This conclusion, after a considerable amount of discussion,‡ was confirmed by Hantzsch,§ who showed that the compound reacted as a pseudo-acid.

---

* Ber., 1894, **27**, 365.   † Ber., 1894, **27**, 361.
  ‡ Bamberger, Ber., 1894, **27**, 2601; 1897, **30**, 1248; Annalen, 1900, **311**, 99; Hantzsch, Ber., 1894, **27**, 1729; 1898, **31**, 177; 1899, **32**, 1722.
  § Ber., 1902, **35**, 258.

CHAPTER XV

DIAZO-COMPOUNDS OF THE ALIPHATIC SERIES *

§ 1. **Preparation.**—The amines of the aliphatic series do not react with nitrous acid as do those of the aromatic series. Only in certain cases is there a departure from the usual reaction of substitution of the amino- by the hydroxyl-group, and then the product has not, as might be expected, a composition similar to that of an aromatic diazo-salt, but the nitrogen atoms are each linked to the aliphatic nucleus, thus—

$$R \diagdown \!\! \begin{array}{c} N \\ \| \\ N \end{array}$$

The reason of this difference in behaviour will be explained in the discussion of the constitution of the aromatic diazo-salts (see p. 167).

The first number of the series was obtained by Schiff and Meissen in 1881, who prepared diazocamphor from camphorimide.† This diazo-compound was also obtained by Angeli by the action of nitrous acid on aminocamphor.‡

The principal worker in this field of research is, however, Curtius, who, a little later,§ succeeded in diazotizing the ethyl ester of glycocoll, or ethyl aminoacetate, a reaction which proceeds in two stages ; the first stage is the formation of the nitrite of the aliphatic amine,

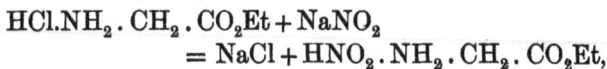

$$HCl.NH_2 . CH_2 . CO_2Et + NaNO_2$$
$$= NaCl + HNO_2 . NH_2 . CH_2 . CO_2Et,$$

and this then loses water with production of the diazo-compound—

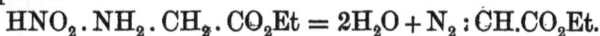

$$HNO_2 . NH_2 . CH_2 . CO_2Et = 2H_2O + N_2 : CH.CO_2Et.$$

---

* Compare also Curtius, *Diazoverbindungen der Fettreihe*, 1886 ; Démètre Vladesco, *Sur les composés diazoïques de la série grasse*, 1891.
† *Gazzetta*, 1881, **11**, 171.
‡ See also Angeli, *Ber.*, 1904, **37**, 2080, footnote.
§ *Ber.*, 1883, **16**, 2230 ; *J. pr. Chem.*, 1888 [ii], **38**, 401.

The free fatty acids do not yield diazo-derivatives, as these are immediately decomposed, so that the esters, amides, &c., must be used. The constitution of diazoacetic ester is proved by (1) the ready substitution of the two atoms of nitrogen by two atoms of iodine, yielding di-iodoacetic ester,

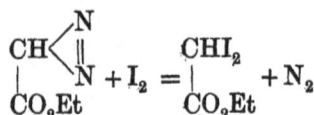

$$CH\begin{matrix} & N \\ & \| \\ & N \\ | & \end{matrix}_{CO_2Et} + I_2 = \begin{matrix} CHI_2 \\ | \\ CO_2Et \end{matrix} + N_2$$

and (2) the reduction to ammonia and glycocoll—

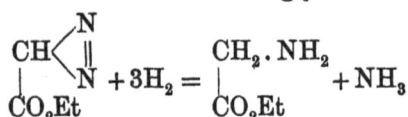

$$CH\begin{matrix} & N \\ & \| \\ & N \\ | & \end{matrix}_{CO_2Et} + 3H_2 = \begin{matrix} CH_2.NH_2 \\ | \\ CO_2Et \end{matrix} + NH_3$$

Similarly diazosuccinic acid yields ammonia and aspartic acid—

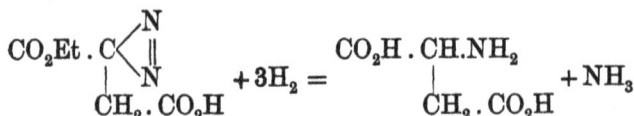

$$CO_2Et.C\begin{matrix} & N \\ & \| \\ & N \\ | & \end{matrix}_{CH_2.CO_2H} + 3H_2 = \begin{matrix} CO_2H.CH.NH_2 \\ | \\ CH_2.CO_2H \end{matrix} + NH_3$$

The preparation of diazoacetic ester is carried out as follows*: Five grams of sodium acetate are dissolved in two litres of water in a ten-litre separating funnel; to this solution one kilo of the finely powdered hydrochloride of ethyl aminoacetate† is added, and then 750 grams of sodium nitrite. The mixture is shaken until the temperature has fallen to about 0°. Five c.c. of ten per cent. sulphuric acid and half a litre of ether are then added and the whole again well shaken. During this period the gradual solution of the still undissolved salts cools the mixture and prevents the reaction from becoming too violent. As soon as the action slackens, the ethereal solution of ethyl diazoacetate is run off, fresh ether added, and ten per cent. sulphuric acid run in from time to time in small quantities until red fumes are evolved. The ethereal solution is then run off, added to that already obtained, washed with small quantities of dilute sodium carbonate solution until the washings assume a deep yellow colour and have an alkaline

* Silberrad, Trans., 1902, 81, 600.
† For the preparation of this compound see Hantzsch and Silberrad, Ber., 1900, 33, 70.

reaction. The ethereal solution is dried by shaking with fused calcium chloride, and freed from ether on the water-bath. The yield amounts to 770 grams, or 94·7 per cent. of the theoretical quantity.

§ **2. Properties of diazoacetic esters.**—The esters of diazo-acetic acid are liquids which solidify at very low temperatures. They are citron-yellow, and possess a characteristic odour. On being warmed to 100° the colour changes to a deep orange, but, on cooling, the original colour reappears. The esters boil without decomposition under very low pressures, and even at the ordinary pressure by quick distillation over the free flame the greater part of the liquid passes over unchanged; the rest decomposes with slight detonation, and forms a thick, white cloud.

The ethyl ester is extraordinarily volatile, and rapidly vaporizes in a vacuum over sulphuric acid.

The esters distil mostly unchanged with steam, the volatility increasing with the weight of the ester radical, whilst the solubility in water at the same time decreases.

The diazo-compounds of the fatty esters are miscible in all proportions with alcohol, ether, benzene, light petroleum, &c.

Methyl diazoacetate, $N_2:CH.CO_2.CH_3$, boils at 129° under a pressure of 721 mm. Its sp. gr. is 1·139 at 21°. It is moderately soluble in water, and has a neutral reaction.

Ethyl diazoacetate, $N_2:CH.CO_2.C_2H_5$, crystallizes in a mixture of ether and solid carbon dioxide to a crystalline mass, which melts at −24°. It boils at 143–144° under 721 mm. pressure, and its sp. gr. is 1·073 at 22°. On gentle warming it takes fire and burns quickly with a luminous flame. It does not explode by concussion, but on adding concentrated sulphuric acid a violent explosion occurs; this also takes place on heating it with certain organic nitro-compounds such as nitroaldehydes.

The ester is sparingly soluble in water, and has a neutral reaction. By heating diazoacetic ester near its boiling-point nitrogen is evolved, and finally fumaric ester remains—

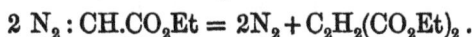

$$2\,N_2:CH.CO_2Et = 2N_2 + C_2H_2(CO_2Et)_2.$$

As an intermediate product in the reaction there is formed azinsuccinic ester *—

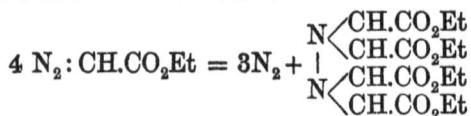

$$4\ N_2:CH.CO_2Et = 3N_2 + \begin{array}{c} N \\ | \\ N \end{array} \begin{array}{l} CH.CO_2Et \\ CH.CO_2Et \\ CH.CO_2Et \\ CH.CO_2Et \end{array}$$

Amyl diazoacetate, $N_2:CH.CO_2.C_5H_{11}$, boils at 160° under a pressure of 721 mm. It is insoluble in water, and has a neutral reaction.

When diazoacetic esters are mixed with aqueous potassium hydroxide or baryta water the corresponding metallic salts are formed. These are stable, however, only in cold, aqueous solution, and Curtius was unable to isolate them or to prepare the free acid by treatment with mineral or organic acids. Thiele, however, by another method, succeeded in preparing the pure sodium salt (see p. 109).

With concentrated aqueous ammonia, diazoacetic esters yield diazoacetamide, $N_2:CH.CO.NH_2$, which crystallizes from warm alcohol or water in large, gold-yellow, transparent prisms. These crystals melt at 114° with copious evolution of gas.

§ **3. Reactions of the aliphatic diazo-compounds.**—The reactions of the fatty diazo-compounds are very similar to those of the aromatic; thus, with water, nitrogen is evolved, and the corresponding hydroxyester produced.

The reaction in the case of ethyl diazoacetate has been quantitatively studied by Fraenkel.† As in the case of the diazo-compounds of the aromatic series the reaction is unimolecular, and the rate is proportional to the concentration of the hydrogen ions, these exerting a catalytic influence.

The presence of neutral salts destroys the regularity of the decomposition and introduces secondary reactions.

The other decompositions of diazoacetic ester are as follows—
With alcohol it yields ethylglycollic ester.
With picric acid it yields picrylglycollic ester.
With benzaldehyde it yields benzoylglycollic ester.

---

* Curtius, *Ber.*, 1885, **18**, 1302.
† *Zeitsch. physikal. Chem.*, 1907, **60**, 202.

With hydrochloric acid it yields monochloroacetic ester.
With iodine it yields di-iodoacetic ester.

A concentrated solution of hydrofluoric acid yields with diazoacetic ester, diglycollic ester—

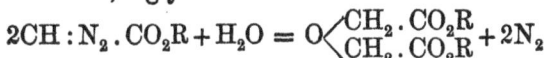

$$2CH:N_2.CO_2R + H_2O = O\!\!<\!\!^{CH_2.CO_2R}_{CH_2.CO_2R} + 2N_2$$

On reduction, diazoacetic esters yield the original amino-compound, a hydrazine being formed as intermediate product, thus—

$$N_2:CH.CO_2R + 2H_2 = NH_2.NH_2.CH_2.CO_2R.$$

A very singular reaction takes place with concentrated aqueous alkalis. An acid is produced having the same composition as diazoacetic acid, but possessing a greater molecular weight. This was considered by its discoverer * to be triazoacetic acid, composed of three molecules of diazoacetic acid, but it was later shown that the substance was really bis-diazoacetic acid—

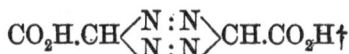

$$CO_2H.CH\!\!<\!\!^{N:N}_{N:N}\!\!>\!\!CH.CO_2H†$$

With potassium sulphite, diazoacetic ester gives the potassium salt of ethyl sulphohydrazimethylenecarboxylic acid—

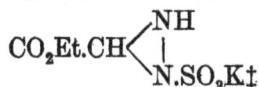

$$CO_2Et.CH\!\!<\!\!^{NH}_{N.SO_3K}‡$$

When ethyl diazoacetate is treated with potassium or sodium ethoxide, the corresponding salt

$$CO_2Et.C\!\!<\!\!^{NK}_{N}$$

of ethyl iso-diazoacetate

$$CO_2Et.C\!\!<\!\!^{NH}_{N}$$

is obtained. This ester is an oil which does not dissolve in

* Curtius, J. pr. Chem., 1889 [ii], 39, 116.
† Hantzsch and Silberrad, Ber., 1900, 33, 58; compare also Curtius, Darapsky, and Müller, Ber., 1907, 40, 1176, 1194.
‡ Von Pechmann, Ber., 1895, 28, 1847.

water, and, unlike the diazoacetate, does not form an additive compound with sulphites.*

The simplest diazo-compound of the aliphatic series, namely, diazomethane,

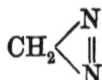

$$CH_2\!\!\left\langle \begin{array}{c} N \\ \| \\ N \end{array} \right.$$

was prepared by von Pechmann in 1894.† This substance, which is a yellow gas, is obtained by warming nitrosoacyl-derivatives of methylamine, $CH_3 . NAc.NO$, with aqueous or alcoholic alkalis. One part by volume of nitrosourethane (from 1 to 5 c.c.), together with 30–50 c.c. of pure ether, and 1·2 parts by volume of 25 per cent. methyl-alcoholic potassium hydroxide, are warmed in a small flask fitted with a reflux condenser on the water-bath.‡ The mixture becomes yellow, and a yellow gas is evolved. The heating is continued until the solution becomes colourless.

The distillate consists of an ethereal solution of diazo-methane, the yield of which is about 50 per cent. of the theory.

Diazomethane is also obtained by the interaction of hydroxylamine and dichloromethylamine—

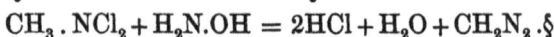

$$CH_3 . NCl_2 + H_2N.OH = 2HCl + H_2O + CH_2N_2 .§$$

The disulphonic acid of diazomethane

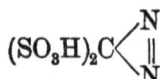

$$(SO_3H)_2C\!\!\left\langle \begin{array}{c} N \\ \| \\ N \end{array} \right.$$

is obtained in a remarkable manner. When potassium cyanide is treated with an aqueous solution of potassium bisulphite in presence of potassium hydroxide, and the product acidified, aminomethanedisulphonic acid results, which, on treatment with nitrous acid, furnishes the corresponding diazo-derivative.‖

---

* Hantzsch and Lehmann, *Ber.*, 1901, **34**, 2506.
† *Ber.*, 1894, **27**, 1888.   ‡ *Ber.*, 1895, **28**, 855.
§ Bamberger and Renauld, *Ber.*, 1895, **28**, 1682 ; compare also Thiele and Meyer, *Ber.*, 1896, **29**, 961.
‖ Von Pechmann and Manck, *Ber.*, 1895, **28**, 2374 ; von Pechmann, *Ber.*, 1896, **29**, 2161.

Phenyldiazomethane is obtained by decomposing potassium benzyldiazo-oxide (p. 110) with water *—

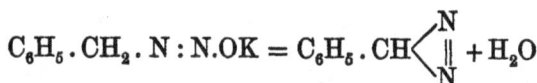

$$C_6H_5.CH_2.N:N.OK = C_6H_5.CH{\diagdown}{\overset{N}{\underset{N}{\parallel}}} + H_2O$$

It is a dark red-brown oil which has only a faint odour, and is slightly volatile. It decomposes when rapidly heated. When distilled under the ordinary pressure it is mostly decomposed with formation of stilbene—

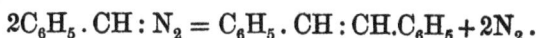

$$2C_6H_5.CH:N_2 = C_6H_5.CH:CH.C_6H_5 + 2N_2 .$$

When warmed with water it yields benzyl alcohol, and in its other reactions it resembles diazomethane.

Another interesting method of obtaining aliphatic diazo-compounds is that due to Traube, which consists in treating the sodium or lead salt of *iso*-nitraminoacetic acid,

$$HO_2N_2.CH_2.CO_2H,$$

with sodium amalgam at 0°. Reduction takes place, and the sodium salt of diazoacetic acid is produced.† A metallic salt of diazoacetic acid was thus isolated in the pure state for the first time.

By treating aminoguanidine nitrate with nitrous acid it was thought that the corresponding diazoguanidine nitrate was formed,‡ but this was later shown to be a derivative of carbaminoiminoazoimide—

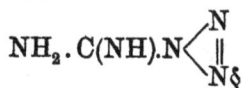

$$NH_2.C(NH).N{\diagdown}{\overset{N}{\underset{N\S}{\parallel}}}$$

§ 4. Metallic diazo-compounds of the aliphatic series.—
The compounds of this class were obtained by Hantzsch and Lehmann ‖ by treating nitrosoalkylurethanes with concentrated potassium hydroxide solution or ethereal potassium ethoxide, thus—

---

* Hantzsch and Lehmann, *Ber.*, 1902, **35**, 897.
† *Ber.*, 1896, **29**, 667.
‡ Thiele, *Annalen*, 1892, **270**, 1; E. P. 2194 of 1892; Thiele and Osborne, *Ber.*, 1897, **30**, 2867; *Annalen*, 1899, **305**, 64.
§ Hantzsch and Vagt, *Annalen*, 1901, **314**, 339.
‖ *Ber.*, 1902, **35**, 897.

$$CH_3 . N\underset{CO_2Et}{\overset{NO}{\diagdown}} \underset{\overset{KOEt}{\longrightarrow}}{\overset{KOH}{\longrightarrow}} \begin{matrix} CH_3 . N : N.OK + H_2O + K_2CO_3 \\ CH_3 . N : N.OK + EtOH + KEtCO_3 \end{matrix}$$

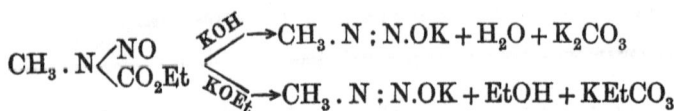

These salts are highly unstable; with water they decompose with explosive violence. The metallic salt obtained from nitrosomethylurethane forms diazomethane, and that derived from nitrosobenzylmethane gives phenyldiazomethane.

Potassium methyldiazo-oxide, $CH_3 . N:N.OK + H_2O$, separates in white crystals when nitrosomethylurethane is gradually added to a very concentrated aqueous solution of potassium hydroxide at $0°$. As excess of alkali is used, the reaction proceeds according to the equation—

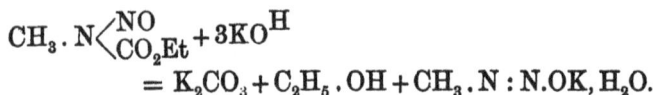

$$CH_3 . N\underset{CO_2Et}{\overset{NO}{\diagdown}} + 3KO^H$$
$$= K_2CO_3 + C_2H_5 . OH + CH_3 . N : N.OK, H_2O.$$

Potassium benzyldiazo-oxide, $C_6H_5 . CH_2 . N:N.OK + H_2O$, is obtained in a similar manner from nitrosobenzylurethane.

On decomposition with water it yields, as chief products, phenyldiazomethane and potassium hydroxide, and as secondary products, benzyl alcohol and nitrogen, thus—

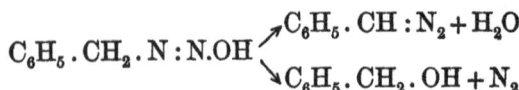

$$C_6H_5 . CH_2 . N : N.OH \diagup\diagdown \begin{matrix} C_6H_5 . CH : N_2 + H_2O \\ C_6H_5 . CH_2 . OH + N_2 \end{matrix}$$

It is found that only esters of α-amino-acids yield diazo-esters; β- and γ-amino-esters, on the other hand, do not form diazo-compounds, and an $\alpha\beta$-diamino-ester therefore yields an α-diazo-β-hydroxy-ester.

Further, only fatty compounds in which the amino-group, carbonyl, and at least one hydrogen atom are attached to the same carbon atom, yield diazo-compounds with nitrous acid;[*] thus, for example, diazoacetophenone is obtained by adding sodium nitrite solution to an aqueous solution of the hydrochloride of aminoacetophenone, and then dropping acetic acid into the cold solution. A solid substance separates, which is

* Curtius and Müller, *Ber.*, 1904, **37**, 1261 ; compare also Angeli, *Ber.*, 1904, **37**, 2080.

washed with sodium carbonate solution and crystallized from light petroleum. Yellow needles are obtained, possessing the constitution—

$$C_6H_5.CO.CH{\Big\langle}\begin{smallmatrix}N\\ \|\\ N\end{smallmatrix} \quad *$$

§ 5. **Diazoamino-compounds of the aliphatic series.**—The simplest representative of the aliphatic diazoamino series, namely, diazoaminomethane, is obtained by the action of magnesium methyl iodide on methylazoimide † and decomposition of the resulting compound with water.‡

Its formation is represented by the equations—

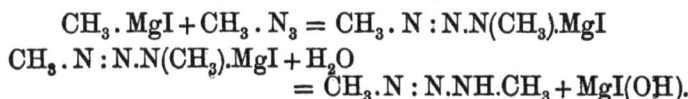

$$CH_3.MgI+CH_3.N_3 = CH_3.N:N.N(CH_3).MgI$$
$$CH_3.N:N.N(CH_3).MgI+H_2O$$
$$= CH_3.N:N.NH.CH_3+MgI(OH).$$

Diazoaminomethane is a colourless liquid, boiling at 92°, which solidifies when immersed in a mixture of solid carbon dioxide and ether; the solid melts at −12°. When heated quickly it explodes, and it decomposes with acids according to the equation—

$$N_3(CH_3)_2H+2HCl = CH_3Cl+N_2+NH_2.CH_3, HCl.$$

* Angeli, *Ber.*, 1893, **26**, 1715 ; *Gazzetta*, 1895, **25**, ii, 494.
† Prepared by treating sodium azoimide with methyl sulphate (Dimroth and Wislicenus, *Ber.*, 1905, **38**, 1573).
‡ Dimroth, *Ber.*, 1906, **39**, 3905.

CHAPTER XVI

CONSTITUTION OF THE DIAZO-COMPOUNDS

ON account of the extraordinary controversy which has raged round the subject of the constitution of the diazo-compounds, it has appeared more advisable to deal with this question separately and more fully than would otherwise have been possible.

As will have been gathered from the account already given, the first question calling for attention is that of the constitution of the diazo-salts, and then naturally follows that of the two classes of isomeric metallic diazo-derivatives.

§ 1. Constitution of the diazo-salts according to Griess.—The first attempt to assign a constitutional formula to a diazo-salt was made by Griess, who gave to diazobenzene nitrate the formula $C_6H_4N_2$, HNO_3.*

Griess considered that in a diazo-compound two atoms of hydrogen of the benzene nucleus were substituted by two atoms of (monoatomic) nitrogen.

In 1859 Wurtz † suggested that each atom of nitrogen was tervalent, and that a bivalent group N_2'' was present.

Erlenmeyer ‡ and Butleroff § developed this idea and gave to diazobenzene the formula

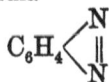

$$C_6H_4\diagdown\begin{matrix}N\\ \| \\ N\end{matrix}$$

which Griess adopted.‖

The idea that two hydrogen atoms of the benzene nucleus were substituted by nitrogen was still present.

Griess also, in the same year, considered that a possible formula for diazobenzene nitrate was $C_6H_4 : N : N, HNO_3$.

---

* *Phil. Trans.*, 1864, **154**, 667.
† *Répert. de Chimie pure*, 1858–9, **1**, 348.
‡ *Zeitsch. f. Chem.*, 1861, 176 ; 1863, 678.
§ *Zeitsch. f. Chem.*, 1863, 511.   ‖ *Ber.*, 1874, **7**, 1618.

About this time Griess discovered that the diazoamino-compound obtained from aniline and bromodiazobenzene nitrate was identical with that prepared from bromoaniline and diazobenzene nitrate, and therefore put forward for diazoaminobenzene the symmetrical formula—

$$C_6H_4 : NH.NH.NH : C_6H_4 .$$

It is interesting to notice in passing that in the aliphatic series the diazo-group is actually united with two valencies of carbon, thus—

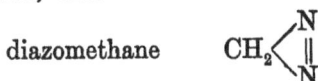

diazomethane $\quad CH_2 \begin{array}{c} \diagup N \\ \| \\ \diagdown N \end{array}$

§ 2. Constitution of diazo-compounds according to Kekulé.

—In 1886 Kekulé advanced the view that diazo-compounds contained the group .N : N., and considered that the behaviour of these compounds was not in accord with Griess's idea that two atoms of hydrogen in the benzene nucleus were displaced;* thus the formation of diazobenzene nitrate proceeded, according to Kekulé, as follows—

$$C_6H_5 . NH_2, HNO_3 \rightarrow C_6H_5 . N : N.NO_3 .$$

Kekulé's opinion that, in diazobenzene nitrate, there were five, and not four, atoms of hydrogen attached to the benzene nucleus was proved by the fact that penta-substituted derivatives of aniline could be converted into the corresponding diazo-salt without suffering any loss of their substituents.†

The formulae which Kekulé introduced were thus—

Free diazobenzene . . .	$C_6H_5 . N : N.OH$
Diazobenzene sulphate . .	$C_6H_5 . N : N.HSO_4$
Diazobenzene platinichloride .	$(C_6H_5 . N : N.Cl)_2PtCl_4$
Potassium salt . . .	$C_6H_5 . N : N.OK$
Diazoaminobenzene . .	$C_6H_5 . N : N.NH.C_6H_5$
Diazobenzene perbromide .	$C_6H_5 . N : NBr, Br_2$
or .	$C_6H_5 . NBr.NBr_2$

Whereas Griess regarded the diazobenzene salts as additive

* *Lehrbuch der organischen Chemie,* II. 717.
† Langfurth and Spiegelberg, *Annalen,* 1878, **191**, 205 ; 1879, **197**, 305 ; compare also ibid., 1874, **174**, 355 ; 1880, **215**, 103.

compounds of diazobenzene and acids, Kekulé looked upon diazobenzene as playing the part of a base analogous to ammonium.

Further, he explained the difference in stability between a diazo-salt such as diazobenzene chloride and an azo-compound

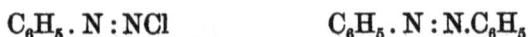

$$C_6H_5 . N : NCl \qquad\qquad C_6H_5 . N : N.C_6H_5$$

as being due to each nitrogen atom being united with a phenyl group in the latter, whilst in the former the union of chlorine and nitrogen rendered the compound similar to chloride of nitrogen. Diazoaminobenzene was formed, according to Kekulé, by the union of the acidic part of the diazo-salt with a hydrogen atom of aniline, and the residues of both uniting thus—

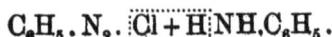

$$C_6H_5 . N_2 . \overline{Cl + H} NH.C_6H_5 .$$

The compound thus was an anilide, and hence was called diazobenzene anilide.

This conception of the constitution of diazoaminobenzene is supported by the fact that, like many hydrazones, it forms metallic compounds, the hydrogen of the NH group being replaceable.

Diazobenzene perbromide was considered by Kekulé to be either an additive compound of diazobenzene bromide with one molecule of bromine (1) or a compound of formula (2)

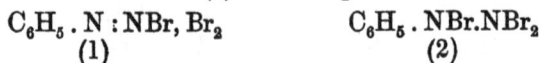

$$C_6H_5 . N : NBr, Br_2 \qquad\qquad C_6H_5 . NBr.NBr_2$$
$$(1) \qquad\qquad\qquad (2)$$

the former of which was regarded as the correct one.

The constitution of diazobenzene imide was correctly written by Kekulé as

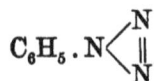

$$C_6H_5 . N{\Large\langle}{\overset{N}{\underset{N}{\|}}}$$

and he pointed out that although free diazobenzene (or diazobenzene hydroxide), $C_6H_5 . N : N.OH$, was very unstable, yet certain other diazo-compounds could exist in the free state, such as, for example, the diazophenols. The reason of this was, that whilst one nitrogen of the diazo-group was attached to the benzene ring, the other was united to the oxygen atom of the phenol group, thus—

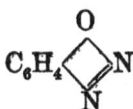

$$C_6H_4 \underset{N}{\overset{O}{\diamondsuit}} N$$

This view was confirmed by the fact that the diazo-derivatives of phenol ethers—for example, anisole—behave as derivatives of diazobenzene and not of diazophenol. Thus salts of nitrodiazoanisole with mineral acids are easily obtainable, whilst those of diazophenol are not; further, free diazoanisole does not exist.

The diazosulphonic acids possessed a similar constitution to the diazophenols; thus diazotized sulphanilic acid was

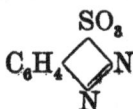

$$C_6H_4 \underset{N}{\overset{SO_3}{\diamondsuit}} N$$

Kekulé also pointed out that the diamines could be divided into three classes according to their behaviour on diazotization, namely, (1) those in which only one amino-group was diazotized, (2) those in which both amino-groups were diazotized, or (3) those in which one amino-group was diazotized and the other took part in the reaction.*

Many examples of these three classes have already been described in the foregoing pages. The difference in stability between the aromatic diazo- and azo-compounds was explained by Kekulé to lie in the fact that whilst both contained the group $C_6H_5 . N :$, in the latter series it was attached to a benzene radical, whilst in the former a halogen or acid radical was united with it.

Kekulé's theory of the constitution of diazo-compounds was universally adopted, and held its own for thirty years, until, in fact, the discovery of the isomeric metallic salts, giving, as it did, an immense impetus to the study of their constitution, led to the abandonment of Kekulé's theory in favour of that of Blomstrand.

* Compare Holleman, *Zeitsch. f. Chem.*, 1865, 557; Hofmann, *Annalen*, 1860, **115**, 251.

§ 3. Constitution of diazo-salts according to Blomstrand.

—An entirely novel view of the constitution of the diazo-salts was published by Blomstrand in 1869.* This chemist explained the formation of diazoaminobenzene and of diazobenzene nitrate in the following way.

In order to obtain a diazo-compound from an amine and one molecule of nitrous acid (HONO), three atoms of hydrogen must be present in order to become replaced by an atom of nitrogen. In the preparation of diazoaminobenzene from an alcoholic solution of aniline and nitrous acid, two molecules of aniline are required to furnish these three hydrogen atoms; consequently a simple diazo-compound is not the final product—

$$\begin{array}{c} C_6H_5.NH\,H \\ C_6H_5.N\,HH \end{array} \rightarrow \begin{array}{c} C_6H_5.NH \\ C_6H_5.N\,N \end{array}$$

If, however, the starting-point is a salt of aniline, which, according to the ammonium theory of Berzelius, is a salt of a substituted ammonium ($C_6H_5.NH_3$), the three necessary hydrogen atoms are now present. Further, in the latter is present a quinquevalent nitrogen atom, whilst in diazoaminobenzene the two atoms of nitrogen are in the tervalent condition.

The formation of diazobenzene nitrate is therefore to be regarded as follows—

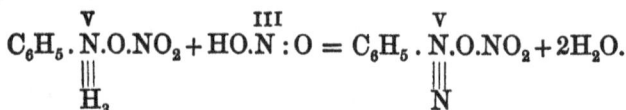

$$\overset{V}{C_6H_5}.\underset{\underset{H_3}{\overset{|||}{}}}{\overset{III}{N}}.O.NO_2 + HO.\overset{V}{N}:O = C_6H_5.\underset{\overset{|||}{N}}{N}.O.NO_2 + 2H_2O.$$

The three atoms of hydrogen attached to the quinquevalent nitrogen atom in aniline nitrate have thus been replaced by a tervalent nitrogen atom.

This theory of the constitution of the diazo-salts was also put forward independently by Strecker in 1871 † and by Erlenmeyer in 1874 ‡ without the knowledge of Blomstrand's paper § or of each other.

* *Chemie der Jetztzeit*, 1869, No. 4, 272.
† *Ber.*, 1871, **4**, 786.   ‡ *Ber.*, 1874, **7**, 1110.
§ The *Chemie der Jetztzeit* seems to be a very obscure publication. No copy exists in the Patent Office library or that of the Chemical Society.

That these two chemists had been anticipated was shown in a paper by Blomstrand in 1875,* who thus established his claim to priority.

Blomstrand pointed out the superiority of his formula to that of Kekulé in that no change in the valency of the aniline-nitrogen was postulated, a change for which there is no justification—

$$\text{Kekulé} \qquad C_6H_5 . \overset{\text{V}}{N}H_3Cl \;\rightarrow\; C_6H_5 . \overset{\text{III}}{N} : NCl$$

$$\text{Blomstrand} \quad C_6H_5 . \overset{\text{V}}{N}H_3Cl \;\rightarrow\; C_6H_5 . \overset{\text{V}}{\underset{\underset{N}{\|\|\|}}{N}}Cl$$

He also explained the instability of the diazo-salts by referring to the unusual replacement of three atoms of hydrogen by one of nitrogen in an ammonium salt.

Blomstrand agreed with Erlenmeyer in adopting the formula $C_6H_5 . NBr : NBr_2$ for diazobenzene perbromide.

In later papers Blomstrand developed his theory more fully in the light of recent work and suggested the names:— 'Azoammonium' compounds for the diazo-salts; 'azo'-compounds for not only the stable compounds such as

$$C_6H_5 . N : N.C_6H_5,$$

but even for potassium diazobenzene sulphonate,

$$C_6H_5 . N : N.SO_3K;$$

'diazo'-compounds for the aliphatic diazo-compounds of Curtius, and 'iso-azo' compounds for the labile isomerides of Hantzsch (see later) and the labile forms of metallic salts and hydroxides.†

Blomstrand regarded the unstable azoammonium compounds as readily undergoing change into the azo-compounds under the influence of reagents such as phenols, &c.; for example—

---

* Ber., 1875, 8, 51.
† Acta Reg. Soc. Physiogr. Lund., 6, 1; J. pr. Chem., 1896 [ii], 53, 169.

$$\underset{N}{\overset{R.N.Cl}{|||}} + C_6H_5.OK = KCl + R.N : N.C_6H_4.OH$$

and $\quad \underset{N}{\overset{R.N.Cl}{|||}} + K_2SO_3 = KCl + R.N : N.SO_3K.$

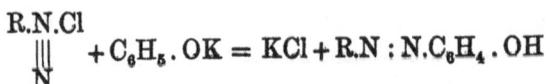

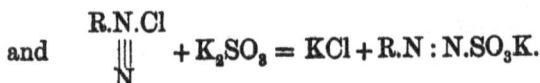

He pointed out that a sharp distinction must be drawn between the quinquevalent salt-forming nitrogen atom in the azoammonium compounds, and the tervalent non-salt-forming nitrogen atom in the azo-compounds.

The chief reason why Blomstrand's theory of the constitution of diazo-salts (azoammonium salts) was not accepted was due to the objection of E. Fischer, who showed that this constitution did not explain the formation of phenylhydrazine, discovered by him in 1875, by simple reduction of a diazo-salt.*

Fischer pointed out that when diazobenzene nitrate was treated with an equimolecular quantity of potassium sulphite, a yellow crystalline salt, $C_6H_5.N_2.SO_3K$, was formed, but when an excess of potassium bisulphite was used, a colourless salt, $C_6H_5.N_2H_2.SO_3K$, was obtained. The latter, potassium phenylhydrazine sulphonate, had already been prepared by Strecker and Römer in 1871.†

The former salt had all the properties of a diazo-compound, and on treatment with zinc dust and acetic acid passed into the latter, which was, therefore, a product of reduction. It had no longer the properties of the diazo-compound but was converted into this by gentle oxidation.

On treatment with hydrochloric acid, the sulphonic acid group was eliminated, and there resulted the hydrochloride of phenylhydrazine. Strecker and Römer formulated these compounds as follows—

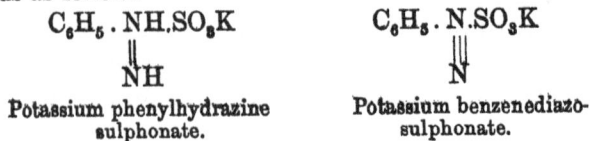

$$\underset{NH}{\overset{C_6H_5.NH.SO_3K}{||}} \qquad\qquad \underset{N}{\overset{C_6H_5.N.SO_3K}{|||}}$$

Potassium phenylhydrazine   Potassium benzenediazo-
sulphonate.   sulphonate.

(The latter formula, it will be noticed, differs from that suggested by Blomstrand.)

* *Ber.*, 1877, 10, 1331.　　　　　　　　† *Ber.*, 1871, 4, 784.

Fischer pointed out that in order to explain the formation of phenylhydrazine from a diazobenzene salt according to the Blomstrand theory, it would be necessary to assume the change of quinquevalent nitrogen into tervalent by the addition of hydrogen, and also the change of tervalent into quinquevalent nitrogen by the withdrawal of hydrogen,

$$\overset{\text{v}}{C_6H_5}.\overset{\text{}}{N}.Cl \underset{\underset{N}{\overset{|||}{}}}{} + 4H \ \rightleftarrows \ C_6H_5.\overset{\text{III v}}{NH.NH_3Cl}$$

III

a procedure which was extremely improbable.

On the other hand, if we assume that no change of valency occurs when phenylhydrazine is formed, we have

$$C_6H_5.\underset{\underset{N}{\overset{|||}{}}}{N}.Cl \ \rightarrow \ C_6H_5.\underset{\underset{NH}{\overset{||}{}}}{NH_3}Cl$$

giving us for free phenylhydrazine the formula—

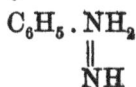

$$C_6H_5.\underset{\underset{NH}{\overset{||}{}}}{NH_2}$$

Fischer proved, however, that this formula could not represent the constitution of phenylhydrazine * in the following way: phenylhydrazine and ethyl bromide unite to form the compound

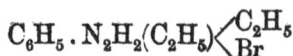

$$C_6H_5.N_2H_2(C_2H_5)\diagdown\diagup{}^{C_2H_5}_{Br}$$

which is also produced by the addition of ethyl bromide to phenylethylhydrazine. As the latter substance, however, is derived from ethylaniline by the substitution of the remaining N-hydrogen atom by the group NH, and thus possesses the constitution

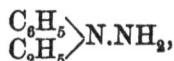

$$\genfrac{}{}{0pt}{}{C_6H_5}{C_2H_5}\diagup N.NH_2,$$

therefore the compound of phenylethylhydrazine and ethyl bromide must contain the complex $> N.NH_2$, and consequently phenylhydrazine itself must be $C_6H_5.NH.NH_2$. Fischer thus

* *Annalen*, 1877, **190**, 67.

concluded that its formation from diazobenzene chloride could only be explained by the aid of Kekulé's formula—

$$C_6H_5 . N : NCl \quad \rightarrow \quad C_6H_5 . NH.NH_3Cl.$$

When the whole question of the constitution of the diazo-compounds was undergoing renewed investigation many years later, Blomstrand explained that the formation of phenyl-hydrazine might be expressed as follows—

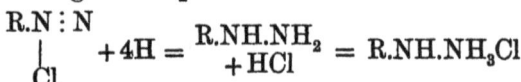

$$\begin{matrix} R.N:N \\ | \\ Cl \end{matrix} + 4H = \begin{matrix} R.NH.NH_2 \\ +HCl \end{matrix} = R.NH.NH_3Cl$$

and pointed out that it was impossible to postulate a double linking between the two nitrogen atoms as shown in the formula

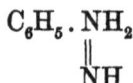

$$C_6H_5 . \underset{NH}{\overset{NH_2}{\|}}$$

when complete reduction had taken place.*

The objections of Fischer, however, as has been said, were taken as final, and it was only in 1895 that the formula of Blomstrand was again adopted.

It seems suitable, at this stage, to postpone further inquiry as to the constitution of the diazo-salts until we have considered a little more fully that of the free diazobenzene, for the two are very closely connected.

§ 4. **Constitution of diazobenzene hydroxide to 1894.**—In chapter xiii we have seen that when diazo-compounds are condensed with various substances with the formation of mixed azo-compounds, the resulting substances were regarded in some cases as true azo-compounds and in others as hydrazones. It was therefore argued by the workers in this field that the original diazo-compound (or rather the hydroxide which might be supposed to be formed) might also be represented by a tautomeric formula—

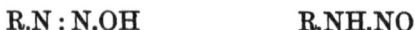

$$R.N : N.OH \qquad\qquad R.NH.NO$$

the former of which would give rise to true azo-compounds, and the latter to the hydrazones.†

* *J. pr. Chem.*, 1896 [ii], **53**, 176.
† V. Meyer, *Ber.*, 1888, **21**, 15; Japp and Klingemann, *Ber.*, 1891, **24**, 2264; von Pechmann, *Ber.*, 1891, **24**, 3255; Bamberger, *Ber.*, 1891, **24**, 3264; 1893, **26**, 495.

It has also been shown that Bamberger's discovery of the diazoic acids lent support to this view * (p. 102).

Further confirmation was adduced by the experiments of von Pechmann † on the action of acetic anhydride on an alkaline solution of diazobenzene. He found that an acetylated nitrosoamine was formed which was identical with the nitrosoamine prepared by O. Fischer ‡ by the action of nitrous acid on acetanilide.

Von Pechmann therefore concluded that primary amines by successive treatment with (1) nitrous acid, (2) acetic anhydride, or (1) acetic anhydride, (2) nitrous acid, yielded the same product, and that free diazobenzene was to be regarded as the anilide of nitrous acid—

OH.NO C_6H_5 . NH.NO

Nitrous acid.   Anilide of nitrous acid.

In confirmation of this view von Pechmann showed that nitrosoanilides (prepared from an anilide and nitrous acid) actually coupled with primary amines and phenols, yielding diazoamino- and hydroxyazo-compounds respectively, the acetyl group being at the same time split off.

The discovery by von Pechmann and Frobenius § that the methyl ether prepared from the silver salt of p-nitrodiazobenzene, to which reference has already been made (p. 98), was isomeric with that obtained from the sodium salt of Schraube and Schmidt, gave emphasis to the view of the tautomeric nature of diazobenzene—

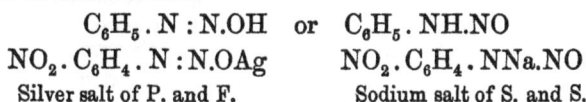

C_6H_5 . N : N.OH   or C_6H_5 . NH.NO

NO_2 . C_6H_4 . N : N.OAg NO_2 . C_6H_4 . NNa.NO

Silver salt of P. and F.   Sodium salt of S. and S.

(See, however, p. 147.)

Further work bearing on this point was immediately published by Bamberger.‖ By treating a β-diazonaphthalene solution with concentrated aqueous sodium hydroxide, an iso-

---

* Hantzsch regarded benzenediazoic acid as

C_6H_5 . N—N.OH
\
O

(*Ber.*, 1894, **27**, 1730).
† *Ber.*, 1894, **27**, 651.
§ *Ber.*, 1894, **27**, 672.

‡ *Ber.*, 1877, **10**, 959.
‖ *Ber.*, 1894, **27**, 679.

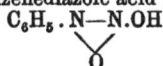

meric substance was obtained which was called β-iso-diazo-naphthalene, $C_{10}H_7 . NH.NO$, and which did not form an azo-compound with alkaline phenols. When, however, it was subjected to the action of a mineral acid, molecular change took place rapidly, and β-diazonaphthalene was re-generated—

$$C_{10}H_7 . NH.NO \longrightarrow C_{10}H_7 . N : N.OH.$$

Bamberger drew the following conclusions as to the mechanism of diazotization. The properties of benzenediazoic acid, especially its transformation to o-nitroaniline, led him to suggest that the first stage in the process of nitration of a primary amine was the formation of a diazoic acid, for he had succeeded in obtaining benzenediazoic acid by the action of nitrogen pentoxide on aniline—

$$C_6H_5 . NH_2 + N_2O_5 \longrightarrow C_6H_5 . NH.NO_2 .$$

Similarly, by the action of nitrous acid on a primary amine, the first product was the nitrosoamine (iso-diazo-compound),

$$C_6H_5 . NH_2 + N_2O_3 \longrightarrow C_6H_5 . NH.NO ;$$

the ordinary form of diazobenzene, $C_6H_5 . N : N.OH$, would then result by molecular change from this.

Bamberger found, in confirmation of this view, that under certain conditions, many primary amines yielded the iso-diazo-compound as first product.*

Further, these isomeric forms of diazobenzene give (according to Bamberger) metallic salts,

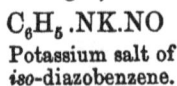

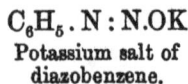

$C_6H_5 .NK.NO$ $C_6H_5 . N : N.OK$

Potassium salt of   Potassium salt of
iso-diazobenzene.   diazobenzene.

of which the iso-salt (like iso-diazobenzene) does not couple with phenols, and is transformed into the normal diazo-salt by mineral acids (see, however, p. 144).

* *Ber.*, 1894, **27**, 1948.

CHAPTER XVII

CONSTITUTION OF THE DIAZO-COMPOUNDS
(*continued*)

§ 1. **Constitution of the diazo-compounds according to Hantzsch.**—An important contribution to the current ideas was next made by Hantzsch.*

He introduced the theory that the constitution of the isomeric diazo-compounds was exactly analogous to that of the isomeric oximes, according to which the former existed as stereoisomeric substances of the general formulae—

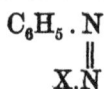   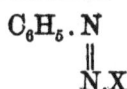

syn.   *anti.*

Hantzsch pointed out that the development of the chemistry of the isomeric diazo-compounds had undergone a precisely similar course to that of the isomeric oximes; to the isomeric diazo-compounds had been assigned the formulae

$$C_6H_5 . N:N.OH \quad \text{and} \quad C_6H_5 . NH.NO,$$

just as, after the discovery of '*iso*-benzaldoxime', its constitution was very generally regarded as being structurally different from that of the normal oxime, thus—

$$C_6H_5 . CH:N.OH$$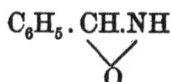

Stable oxime.   Labile oxime.

He showed that the formulae advocated by Schraube and Schmidt, and confirmed by Bamberger, for the metallic diazo-compounds—for example,

$$C_6H_5 . NK.NO \qquad C_6H_5 . N:N.OK$$

iso-salt.   Normal salt.

—required that, in the change from normal to *iso*-salt, the

* *Ber.*, 1894, **27**, 1702.

potassium should wander from the oxygen, for which it has an enormous affinity, to the nitrogen which has little attraction for it. As this transformation takes place, in the case of p-nitrodiazobenzene, at the ordinary temperature and in aqueous solution, the current theory of its mechanism could not be accepted. Further, all substances which exhibit tautomerism in their salts, such as nitrous, cyanic, hydrocyanic, and sulphurous acids, would also have to be considered as displaying structural isomerism in these, but no structural isomeric salts had been found the origin of which isomerism lay in a dissociable group (H or Me) which could alter its position in the molecule. Hence Hantzsch concluded that such isomerism must be steric and not structural. The fact that structurally isomeric alkyl derivatives had been obtained could not be used as a proof of the structural difference of the original substances; Schraube and Schmidt had concluded that because an N-ether (I) was formed from the *iso*-diazobenzene salts and alkyl iodide the original salt had the composition (II)—

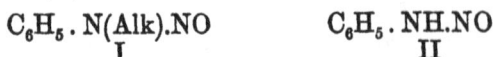

$$C_6H_5 . N(Alk).NO \qquad\qquad C_6H_5 . NH.NO$$
$$\text{I} \qquad\qquad\qquad\qquad \text{II}$$

but this could not be maintained, for if the *iso*-diazobenzene salts were nitrosoamines, their alkyl derivatives ought, by alkaline hydrolysis, to yield the corresponding nitrosoamines ; thus $C_6H_5 . NAlk.NO$ should give $C_6H_5.NH.NO$, but Bamberger had shown that the normal metallic diazo-salt and not the *iso*-salt was formed in each case.*

The production of an N-ether from the potassium salt, and an O-ether from the silver salt (p. 121) of p-nitrodiazobenzene, found an analogy in the case of the oximes.

Hantzsch, therefore, was of the opinion that there was no proof of the structural isomerism of the free phenylnitrosoamines with the true diazo-compounds, and stated that if a derivative of diazobenzene in which the dissociable hydrogen or metallic atom was replaced by a non-dissociable group should exist in two isomeric forms, of which the one showed

---

* Hantzsch himself showed later, however (see p. 147), that in this reaction the *iso*-salts are actually produced.

the reactions of a true diazo-compound (for example, coupling with β-naphthol) and the other those of the *iso*-diazo-compounds, such isomerides would be structurally identical, and their difference would be due to stereoisomerism.

As an example of such isomerism, Hantzsch * described a series of new diazoamino-compounds which he regarded as stereoisomeric with those already known; these were, however, soon shown by Bamberger † not to have any existence, as the substances described by Hantzsch were identical with the bisdiazobenzeneanilides of von Pechmann.‡

A second example was given, namely, a new (*syn*) form of potassium benzenediazosulphonate which, as will be shown, gave rise to a long controversy as to its nature.

For the purpose of determining which diazo-compounds belong to the *syn*- and which to the *anti*-series, Hantzsch took for example those compounds which were considered to form anhydrides, such as diazosulphanilic acid, diazophenol, &c. At that time these were supposed to have the constitution—

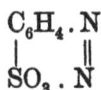

$$C_6H_4 . N$$
$$| \quad \|$$
$$SO_3 . N$$

Anhydride of diazobenzenesulphonic acid.

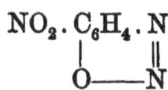

$$NO_2 . C_6H_4 . N$$
$$| \quad \|$$
$$O———N$$

Nitrodiazophenol.

If we now imagine the ring to be broken by addition of

---

* *Ber.*, 1894, **27**, 1857. † *Ber.*, 1894, **27**, 2596.

‡ *Ber.*, 1894, **27**, 703. This is an exceedingly good example of the importance of carrying out complete and exhaustive analyses in organic research. For 'benzene-*syn*-diazoanilide'

$$C_6H_5 . N$$
$$\|$$
$$C_6H_5 . NH . N$$

Hantzsch gave the following numbers:—

Found.: C = 72·7, H = 6.0, N = 20·8,

$C_{12}H_{11}N_3$ requires C = 73·1, H = 5·6, N = 21·3.

The compound was really bisdiazobenzeneanilide,

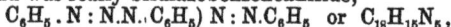

$$C_6H_5 . N : N.N.(C_6H_5) N : N.C_6H_5 \text{ or } C_{18}H_{15}N_5,$$

which requires C = 71·8, H = 5·0, N = 23·2, for which von Pechmann found C = 71·8, 72·3; H = 5·0, 5·8; N = 23·5; and Bamberger, N = 23·2, 23·17 per cent.

For '*p*-toluene-*syn*-diazotoluide' Hantzsch gave—

Found.: N = 19·5,

$C_{14}H_{15}N_3$ requires N = 18·7,

whereas the substance really possessed the formula $C_{21}H_{21}N_5$, which requires N = 20·4 per cent.

water, salt-formation, &c., it is reasonable to suppose that the group attached to the nitrogen atom should retain its position, thus—

$$\begin{array}{ccc} C_6H_4 . N & & C_6H_4\text{———}N \\ | \quad \| & \text{gives} & | \qquad \| \\ SO_3 . N & & SO_3H \ HO.N \end{array}$$

but as all these cyclic compounds readily couple with β-naphthol, it is to be concluded that ordinary diazo-compounds have a similar configuration, or, in other words, those diazo-compounds which combine readily with β-naphthol belong to the *syn*-series, whilst those combining with difficulty or not at all are *anti*-compounds.*

Hantzsch explained the fact that many diazo-compounds, when left for some time in alkaline solution, lose their power of coupling with β-naphthol by saying that the *syn*-compound (alkali-labile) was transformed into the *anti*-compound (alkali-stable) as follows—

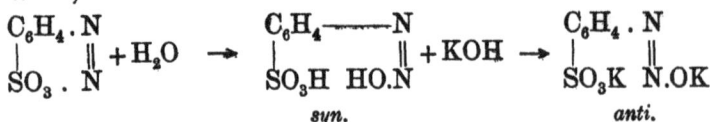

$$\begin{array}{ccccccc} C_6H_4 . N & & C_6H_4\text{———}N & & C_6H_4 . N \\ | \quad \| + H_2O & \rightarrow & | \qquad \| + KOH & \rightarrow & | \quad \| \\ SO_3 . N & & SO_3H \ HO.N & & SO_3K \ N.OK \\ & & syn. & & anti. \end{array}$$

Another method of determining the configuration of the isomeric diazo-compounds was drawn from the analogy to the oximes. In this class of compounds intramolecular change proceeds only in the case of the *syn*-compounds, so that only the diazo-compounds belonging to the same series could decompose according to the equation—

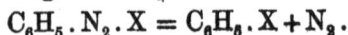

$$C_6H_5 . N_2 . X = C_6H_5 . X + N_2 .$$

The diazo-compounds which correspond to this condition

---

* Hantzsch at a later date accepted the formulae

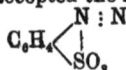

$$C_6H_4 \begin{array}{c} N : N \\ | \\ SO_3 \end{array}$$

and $O : C_6H_4(NO_2) : N_2$ for these compounds, so that this particular argument cannot be maintained. This constitution of the quinonediazides (diazophenols) was adduced by L. Wolff (*Annalen*, 1900, 312, 119 et seq.) from the fact that compounds containing undoubtedly the grouping

$$N \begin{array}{c} O.C. \\ \diagdown \ \| \\ N.C. \end{array}$$

had properties quite different from the former.

are the normal compounds, so these were to be regarded, according to Hantzsch, as *syn*-compounds.

The decomposition thus takes place as follows:—

$$\begin{matrix} C_6H_5 \,|\,.\,N \\ \| \\ X\downarrow.\,N \end{matrix} \quad \rightarrow \quad \begin{matrix} C_6H_5 \\ | \\ X \end{matrix} + N_2$$

On the other hand, the *anti*-compounds have a tendency to decompose into two residues, each containing a nitrogen atom, thus—

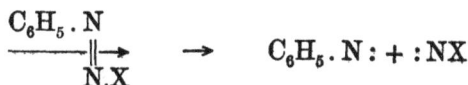

$$\begin{matrix} C_6H_5.\,N \\ \overline{}\|\rightarrow \\ N.X \end{matrix} \quad \rightarrow \quad C_6H_5.\,N : + : NX$$

In this way Hantzsch explained the formation of nitrosobenzene by the oxidation of the *iso*-diazo-compounds—

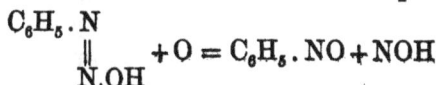

$$\begin{matrix} C_6H_5.\,N \\ \| \\ N.OH \end{matrix} + O = C_6H_5.\,NO + NOH$$

A third method of distinguishing between the *syn-* and *anti*-compounds was to be found in their difference in explosibility, the normal or *syn*-diazo-compounds being much more explosive than the *anti*-compounds.

It is to be mentioned here that Hantzsch regarded diazobenzene chloride as a *syn*-compound,

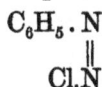

$$\begin{matrix} C_6H_5.\,N \\ \| \\ Cl.N \end{matrix}$$

but soon adopted another view (see p. 133).

Hantzsch's main conclusions were, therefore, that the ordinary normal diazo-compounds were *syn*-diazo-compounds, and the so-called *iso*-diazo-compounds (nitrosoamine formula) were *anti*-diazo-compounds. Bamberger,[*] in criticizing these views of Hantzsch, denied that it was possible to draw a parallel between the stereoisomeric oximes and the isomeric diazo-compounds. He based his objections on the very great difference between the latter, a difference which was greatly in contrast to the very small one existing between the oximes, and which could not be due to stereoisomerism. Moreover,

* *Ber.*, 1894, **27**, 2582.

the isomeric oximes gave corresponding isomeric ethers, but from the diazo- and *iso*-diazo-silver salts only the one normal ether was obtained with methyl iodide.

He maintained that the difference between the normal and the *iso*-diazo-compounds was best explained by the presence of the labile hydrogen atom (*)—

$$C_6H_5 . N : N.OH* \qquad\qquad C_6H_5 . NH*.NO$$

  Normal. *iso.*

Further, the objection of Hantzsch that in the normal and *iso*-potassium benzenediazo-oxides

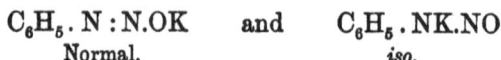

$$C_6H_5 . N : N.OK \qquad \text{and} \qquad C_6H_5 . NK.NO$$

  Normal. *iso.*

it was unlikely that the potassium atom should leave the oxygen and become attached to the nitrogen atom was met by Bamberger by the reminder that many compounds containing the imino-group—for example, azoimide and benziminazole—readily dissolve in alkali, and he pointed out, also, that the positive group $C_6H_5 . N_2$ in the salt $C_6H_5 . N_2 . OK$ greatly lessened the affinity of potassium for the oxygen atom, just as in potassium oxide, K.O.K., one potassium atom is easily removed.

As was briefly indicated above (p. 124), it was obviously of great importance in connexion with Hantzsch's stereochemical theory to adduce evidence of the existence of isomerides other than those capable of being explained by the presence of a labile hydrogen atom which could give rise to isomerism as shown by the formulae—

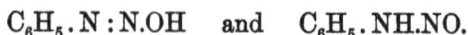

$$C_6H_5 . N : N.OH \qquad \text{and} \qquad C_6H_5 . NH.NO.$$

Hantzsch's first example of stereoisomeric diazoaminobenzenes was, as already mentioned, shown to be based on an error, and considerable interest, therefore, was attached to the other example of stereoisomeric benzenediazosulphonates.

Isomeric benzenediazosulphonates.†—It has been already explained (p. 55) that by treating diazobenzene chloride with neutral potassium sulphite, potassium benzenediazosulphonate is formed ('Strecker's salt'). In 1894 Hantzsch ‡ described a

---

† These must not be confused with the salts of diazobenzenesulphonic acid prepared by diazotizing sulphanilic acid.

‡ *Ber.*, 1894, **27**, 1726.

new isomeride which he obtained by pouring diazobenzene nitrate solution into an ice-cold solution of neutral potassium sulphite containing an excess of potassium carbonate. Orange plates separated which were readily soluble in water and contained one molecule of water of crystallization. The substance was very unstable, and coupled readily with β-naphthol, and Hantzsch assigned to it the *syn*-configuration—

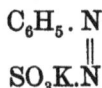

$$C_6H_5 . N$$
$$\|$$
$$SO_3K.N$$

The clear dark-yellow aqueous solution of this new substance gradually became paler on standing, and crystals of the Strecker salt separated out. This was much more stable than the *syn*-compound, and gave no colour reaction at all with alkaline phenol solution. It was, therefore, the *anti*-compound.

Bamberger considered * that the isomeric benzenediazo-sulphonates of Hantzsch

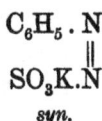

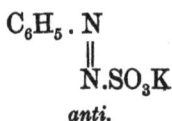

$$C_6H_5 . N \qquad C_6H_5 . N$$
$$\| \qquad \|$$
$$SO_3K.N \qquad N.SO_3K$$
syn. *anti.*

might just as well be represented as

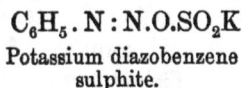

$$C_6H_5 . N : N.O.SO_2K \qquad \text{and} \qquad C_6H_5 . N : N.SO_3K$$
Potassium diazobenzene sulphite. Potassium benzenediazo-sulphonate.

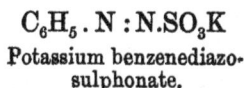

in view of the fact that Hantzsch's new salt gave all the reactions of a sulphite, and he maintained that Hantzsch had not proved that the two were stereoisomerides.

He showed also that Strecker's salt, on acidification, did not pass into the diazo-salt as did *iso*-diazo-compounds.

The same view of the constitution of these salts was expressed by Claus.†

In reply to this criticism, Hantzsch ‡ showed that both modifications gave the same ions, $(C_6H_5 . N_2SO_3)$ and K, in solution, so that they must both have the structure

* *Ber.*, 1894, **27**, 2586, 2930.   † *J. pr. Chem.*, 1894 [ii], **50**, 239.
‡ *Ber.*, 1894, **27**, 2099, 3527.

$C_6H_5 . N_2 . SO_3K$; he pointed out, further, that the sulphite reaction showed only that the new salt decomposed readily with separation of sulphurous acid, just as does the compound $Hg(SO_3K)_2$, which in aqueous solution gives the three ions—

$$Hg(SO_3)_2, K, K.$$

He maintained that a diazo-sulphite, $C_6H_5 . N_2 . O.SO_2K$, must give three ions in solution and not two, although Ostwald had informed Bamberger * that only two ions, namely,

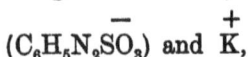

$$(\overset{-}{C_6H_5N_2SO_3}) \text{ and } \overset{+}{K},$$

were to be expected according to analogy.

Up to this point we may summarize the evidence as proving the existence of two isomeric compounds, $C_6H_5 . N_2 . (SO_3K)$. As we have seen, Hantzsch supposed that only one constitutional formula was possible for these, and that the substances were therefore stereoisomeric. Other formulae have, however, been advocated, but the discussion of these must be postponed until the constitution of the diazo-salts (chlorides, &c.) have been more fully dealt with (p. 140).

Isomeric diazo-cyanides.—In 1895 Hantzsch and Schultze succeeded in preparing isomeric diazo-cyanides by the action of potassium cyanide on p-chloro- and p-nitro-diazobenzene chloride solution.†

A little more than the theoretical quantity of potassium cyanide is added to the hydrochloric acid solution of the diazo-salt, and care must be taken that sufficient hydrochloric acid is present to ensure an acid reaction at the end of the operation.

At a low temperature (below −5°) the primary, *syn*, or labile compound is obtained, whilst at higher temperatures the secondary, *anti*, or stable compound is produced.

Both compounds are coloured, crystalline, and almost insoluble in water, and the labile variety passes into the stable form slowly in the solid state but quickly in alcoholic solution. The labile isomerides couple with β-naphthol, and are explosive, whilst the stable do not possess these properties.

The two compounds behave very differently when treated

* *Ber.*, 1894, 27, 2934.   † *Ber.*, 1895, 28, 666.

with copper powder; the *syn*-compounds derived from both
p-chloro- and *p*-nitroaniline yield *p*-chlorobenzonitrile and
p-nitrobenzonitrile respectively, whilst the *anti*-compounds
are entirely without action. Further, the *syn*-compounds
yield azo-dyestuffs with R salt, but the *anti*-compounds
do not. These are substantial chemical differences in the
behaviour of these substances which are not usually met
with in stereoisomeric compounds (see p. 127).

Hantzsch regarded these differences, however, as merely
showing that the one compound was more stable than the
other, and considered the existence of these substances to be
a proof of his stereochemical theory, formulating them as

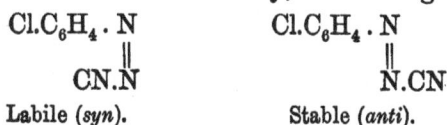

$$\begin{array}{cc} \mathrm{Cl.C_6H_4 . N} & \mathrm{Cl.C_6H_4 . N} \\ \| & \| \\ \mathrm{CN.N} & \mathrm{N.CN} \end{array}$$

Labile (syn).   Stable (anti).

As in the case of the isomeric diazosulphonates, we have
first to discuss the later development of the theory of the
constitution of the diazo-salts with mineral acids before
describing the criticism to which these two series of
isomerides have been subjected.

CHAPTER XVIII

CONSTITUTION OF THE DIAZO-SALTS AFTER 1894

§ 1. **Constitution of the diazo-compounds according to Bamberger.**—In 1894 Bamberger * stated that, in the diazo-salts, the radical (C_6H_5 . N_2) is strongly positive, and that even negatively substituted diazo-salts, such as bromo- and nitro-diazobenzene nitrates, showed a neutral action towards litmus or Congo-red, and were not hydrolytically dissociated in solution. He suggested, therefore, that the formula

$$C_6H_5 . N \vdots N . X$$

was worth consideration. Shortly afterwards,† he pointed out that there were no compounds known in which tervalent nitrogen was combined with a negative complex, such as NO_3, Cl, &c., to form a salt. Hence he concluded that the nitrogen atom united with such groups in the diazo-salts must be quinquevalent, and therefore that Kekulé's formula,

$$C_6H_5 . N : NCl,$$

which was commonly accepted, could not be correct.

He adopted instead of this, or the one just referred to, the old formula suggested by Blomstrand (p. 116), $C_6H_5 . NCl : N$, for the diazo-salts with mineral acids.‡

When the negative group was withdrawn from this (by formation of the hydroxide, for example) the nitrogen atom to which it was attached became tervalent, thus—

$$C_6H_5 . N(OH) : N \quad \longrightarrow \quad C_6H_5 . N : N.OH.$$
Diazobenzene.   *iso*-diazobenzene.

It will be noticed that Bamberger here proposed a formula for the *iso*-compounds differing from the nitrosoamine formula, and a convincing proof of the correctness of this formula for the *iso*-diazo-compounds was apparently given by the discovery

---

* *Ber.*, 1894, **27**, 3417.   † *Ber.*, 1895, **28**, 242.
‡ *Ber.*, 1895, **28**, 444.

that these were formed by the action of hydroxylamine on the nitroso-compounds, thus—

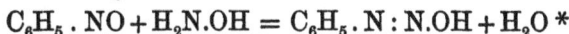

$$C_6H_5 . NO + H_2N.OH = C_6H_5 . N : N.OH + H_2O *$$

but Hantzsch † showed that, in reality, the normal compound was formed as follows—

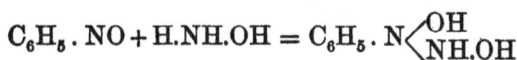

$$C_6H_5 . NO + H.NH.OH = C_6H_5 . N{\overset{OH}{\underset{NH.OH}{<}}}$$

$$\begin{array}{ccc} C_6H_5 . N.OH & & C_6H_5 . N \\ | & \rightarrow & \| \ +H_2O \ \ddagger \\ HO.N.H & & HO.N \end{array}$$

Bamberger's view of the constitution of the diazobenzene salts was not at first accepted by Hantzsch,§ who maintained that diazobenzene chloride in the dry state possessed the constitution

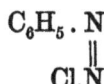

$$\begin{array}{c} C_6H_5 . N \\ \| \\ Cl.N \end{array}$$

and when dissolved in water was to be regarded as the hydrochloride of *syn*-diazobenzene hydroxide—

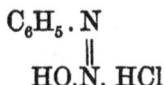

$$\begin{array}{c} C_6H_5 . N \\ \| \\ HO.N, \ HCl \end{array}$$

Very soon, however, Hantzsch gave up the latter idea and adopted Blomstrand's formula for the diazo-salts. He preferred also to call these ' **diazonium** ' salts, from their analogy to the ammonium salts.‖

Bamberger had shown that the diazo-salts had, like the alkali salts, a neutral reaction in solution. That the diazo-salts are electrolytically dissociated in solution had indeed been demonstrated by Goldschmidt in 1890, who also found that they form two ions;¶ and Hantzsch now made a careful comparison of the electrical conductivities of various diazo-salts and salts of the alkali metals, and was able to show that the degree of ionization is about the same in the two cases;

* Bamberger, *Ber.*, 1895, **28**, 1218.   † *Ber.*, 1905, **38**, 2056.
‡ Compare also Angeli, *Ber.*, 1904, **37**, 2390.
§ *Ber.*, 1895, **28**, 676.   ‖ *Ber.*, 1895, **28**, 1734.
¶ *Ber.*, 1890, **23**, 3220.

that is to say, diazonium salts are dissociated almost to the same extent as the corresponding potassium or ammonium salts.

Further analogy was illustrated by Hantzsch's discovery of various double salts of the diazo-salts with cobalt nitrite, mercuric chloride, and mercuric cyanide.

Having thus developed the idea of 'diazonium' as a complex alkali metal, Hantzsch agreed with Bamberger in regarding the salts of 'diazonium' with acids as possessing the constitution which had been attributed to them by Blomstrand, Erlenmeyer, and Strecker, namely, $C_6H_5 . NCl : N$.

The metallic salts, cyanides, sulphonates, &c., belonged to the ordinary 'diazo' form; for example, $C_6H_5 . N:N.OK$ could, as we have seen, exist in two stereoisomeric modifications.

The cyclic diazo-compounds were divided into two groups. On the one hand, Hantzsch assigned to the diazo-compound prepared from sulphanilic acid the formula which had been already given to it by Strecker, namely,

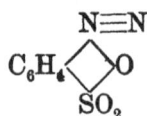

$$C_6H_4 \underset{SO_2}{\overset{N \equiv N}{\diamondsuit}} O$$

and the diazo-phenols and naphthols, which are anhydrides, he regarded as possessing the formulae

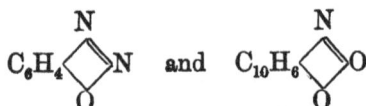

$$C_6H_4 \underset{O}{\overset{N}{\diamondsuit}} N \quad \text{and} \quad C_{10}H_6 \underset{O}{\overset{N}{\diamondsuit}} O$$

respectively.

To this pronouncement of Hantzsch, Bamberger replied that the diazonium radical is not directly comparable with an alkali metal, as its behaviour varies with the substituting group in the aromatic nucleus. Moreover, the electrical conductivity experiments of Goldschmidt and Hantzsch were not a trustworthy basis upon which to speculate as to the nature of diazonium.*

* *Ber.*, 1896, **29**, 446, 564, 608.

Hantzsch now elaborated his arguments in favour of regarding the diazonium radical as a 'compound alkali metal' of the same strength as ammonium or potassium. He pointed out that the alkali salts of all strong monobasic acids, such as hydrochloric, sulphuric, &c., all undergo very extensive electrolytic dissociation in solutions of moderate dilution, but are not hydrolytically dissociated. The dissociation and molecular conductivity are only very slightly increased by further dilution, and the increase ceases at a point of very moderate dilution; the salts of silver and thallium behave similarly, as do also salts of complex ammonium bases, such as mono-, di-, tri-, and tetra-alkylammonium and phenyltrimethylammonium, but not phenylammonium, the ion of the aniline salts. Now the diazonium salts behave in exactly a similar manner,* so that there was strong presumptive evidence that diazonium was constituted similarly to the complex ammoniums.†

§ 2. **Relation between diazonium compounds and normal or *syn*-diazo-compounds.**—Hantzsch's theory that the *syn*-diazo-compounds are those which 'couple' with phenols, &c., to yield azo-dyestuffs led him to explain that these *syn*-compounds were formed as intermediate products in the ordinary reactions of the diazo-salts. The coupling process was thus represented—

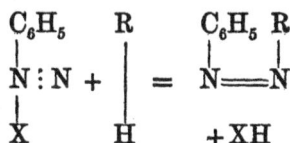

$$
\begin{array}{ccc}
C_6H_5 & R & C_6H_5\ R \\
| & | & |\quad| \\
N:N + & | & = \quad N\!=\!N \\
| & | & \\
X & H & +XH
\end{array}
$$

The formation of the diazo-metallic salts, &c., was expressed as follows—

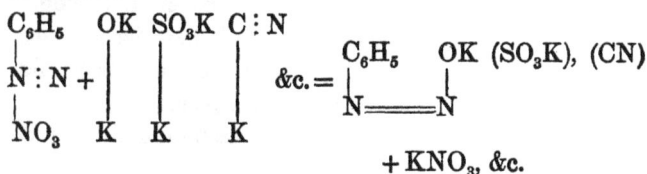

$$
\begin{array}{cccc}
C_6H_5 & OK\ SO_3K\ C:N & & C_6H_5\quad OK\ (SO_3K),\ (CN) \\
| & |\quad|\quad| & &c. = & |\qquad| \\
N:N + & |\quad|\quad| & & N\!=\!N \\
| & |\quad|\quad| & & \\
NO_3 & K\quad K\quad K & & +KNO_3,\ \&c.
\end{array}
$$

---

* *Ber.*, 1895, **28**, 1737.   † *Ber.*, 1895, **28**, 1740; 1898, **31**, 1612.

and the decomposition of the diazonium salts was explained
similarly—

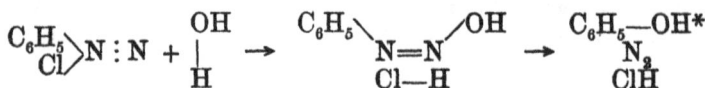

$$\underset{Cl}{\overset{C_6H_5}{>}}N:N + \underset{H}{\overset{OH}{|}} \rightarrow \underset{Cl-H}{\overset{C_6H_5}{\diagdown}}N=N\overset{OH}{\diagup} \rightarrow \underset{ClH}{\overset{C_6H_5-OH*}{N_2}}$$

Diazosulphanilic acid, when treated with one molecule of
alkali, passes into the mono- and with two molecules of alkali
into the di-alkali salt, thus—

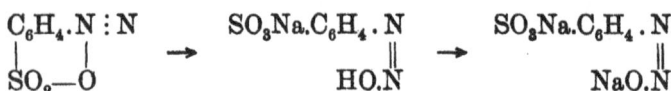

$$\underset{SO_2-O}{\overset{C_6H_4.N:N}{|\quad|}} \rightarrow \underset{HO.N}{\overset{SO_3Na.C_6H_4.N}{\|}} \rightarrow \underset{NaO.N}{\overset{SO_3Na.C_6H_4.N}{\|}}$$

This view of the intermediate formation of *syn*-diazo-com-
pounds in the reactions of the diazonium salts received confirmation from the experiments of Hantzsch and Gerilowski,[†]
who showed that whereas free diazosulphanilic acid is fairly
stable in aqueous solution, the primary alkali salt obtained
by the action of one molecule of alkali, under the same conditions, loses practically all its nitrogen, thus—

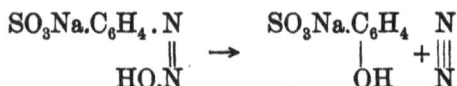

$$\underset{HO.N}{\overset{SO_3Na.C_6H_4.N}{\|}} \rightarrow \underset{OH}{\overset{SO_3Na.C_6H_4}{|}} \underset{N}{\overset{N}{\|}}$$

It is evident that the nature of the decomposition of the
diazo-salts is different from that of the ammonium salts, for
if the two resembled each other, one would expect that as the
group attached to the diazonium radical becomes less negative the stability of the salt would decrease just as
NH_4Cl, $(NH_4)_2CO_3$, and $NH_4.OH$ decrease in stability.

This is, however, not the case, as solutions of diazonium
carbonates are comparatively stable. Moreover, one would
expect also that diazonium halogen salts, if they decomposed
in aqueous solution analogously to the ammonium salts, would
yield halogen-substituted benzenes and not phenols.

Hantzsch explained this difference by assuming the inter-

---

* *Ber.*, 1895, **28**, 1734; 1900, **33**, 2517.
† *Ber.*, 1896, **29**, 1063.

mediate formation of *syn*-diazobenzene hydroxide, which then could decompose into nitrogen and phenol—

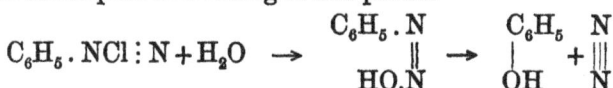

$$C_6H_5 . NCl : N + H_2O \rightarrow \begin{array}{c} C_6H_5 . N \\ \parallel \\ HO.N \end{array} \rightarrow \begin{array}{c} C_6H_5 \\ | \\ OH \end{array} + \begin{array}{c} N \\ \parallel \\ N \end{array}$$

The action of alcohol was explained in this way:—

(1) Formation of ethers—

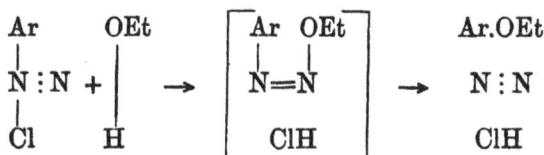

$$\begin{array}{cc} Ar & OEt \\ | & | \\ N:N & + \\ | & | \\ Cl & H \end{array} \rightarrow \left[\begin{array}{cc} Ar & OEt \\ | & | \\ N=N \\ | \\ ClH \end{array} \right] \rightarrow \begin{array}{c} Ar.OEt \\ \\ N:N \\ \\ ClH \end{array}$$

(2) Formation of hydrocarbons—

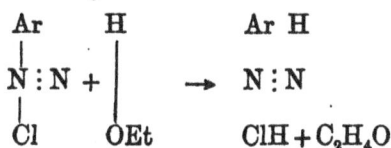

$$\begin{array}{cc} Ar & H \\ | & | \\ N:N & + \\ | & | \\ Cl & OEt \end{array} \rightarrow \begin{array}{c} Ar \; H \\ \\ N:N \\ \\ ClH + C_2H_4O \end{array}$$

§ 3. Double salts of diazonium halides and metallic salts.

—Double salts of diazo-halides with platinum and gold chlorides have been known since the days of Griess and a large number of others have since been prepared.

Hantzsch has further shown * that two kinds exist, namely, colourless, stable, diazonium halogen double salts and coloured, labile, *syn*-diazo-halogen double salts.

For example we have—

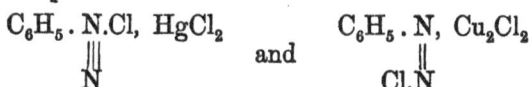

$$C_6H_5 . N.Cl, \; HgCl_2 \atop \parallel \atop N \qquad \text{and} \qquad C_6H_5 . N, \; Cu_2Cl_2 \atop \parallel \atop Cl.N$$

Certain of the diazo-halides can also unite with halogen acids to form additive compounds of formulae

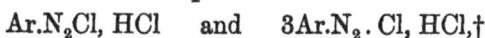

$$Ar.N_2Cl, HCl \quad \text{and} \quad 3Ar.N_2 . Cl, HCl,\dagger$$

the constitution of the former being represented by Hantzsch as

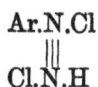

$$Ar.N.Cl \atop \parallel \atop Cl.N.H$$

* *Ber.*, 1895, **28**, 1736.   † Hantzsch, *Ber.*, 1897, **30**, 1153.

and the latter being regarded as compounds of two molecules of the neutral and one of the acid salt.

§ 4. **Diazonium halides and** *syn*-**diazo-halides.**—Hantzsch found that all diazo-bromides, iodides, and thiocyanates which belong to the same series as colourless diazonium chlorides and salts of oxygen acids are coloured when in the solid state; with increase of colour is noticed an increase in the explosibility. Thus we have

Diazo-chloride.	**Diazo-bromide.**	**Diazo-thiocyanate.**
Colourless.	Slightly coloured.	Strongly coloured.
Hardly explosive.	Slightly explosive.	Very explosive.

Diazo-iodide.
Intensely coloured.
Extremely explosive.

As diazonium salts should be, like the corresponding alkali and ammonium salts, colourless, and as, on the other hand, *syn*-diazo-halides, from analogy to the coloured *syn*-diazo-cyanides, should be coloured, and also, as being compounds of the type of nitrogen iodide, would be expected to be explosive, Hantzsch * concluded that the properties of the above series of compounds were only to be explained by the assumption that they consist of an equilibrium mixture † of colourless diazonium halides and coloured *syn*-diazo-halides, thus—

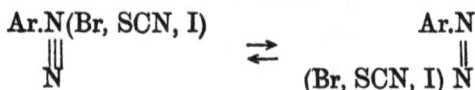

$$\text{Ar.N(Br, SCN, I)} \atop \underset{N}{\overset{|||}{}} \quad \rightleftharpoons \quad \text{(Br, SCN, I)} \; \underset{}{\overset{Ar.N}{\overset{||}{}N}}$$

the chlorides belonging entirely to the diazonium series, and the cyanides to the *syn*-series.

The proportion of *syn*-diazo-compound in the mixture becomes less with a lowered temperature,‡ for at − 60° many diazo-halides are nearly colourless and become more intensely coloured with rise of temperature. In the colourless aqueous solutions, of course, the *syn*-compound has become entirely transformed into the diazonium isomeride.

§ 5. **Diazonium perhalides.**—Griess found that two bromine atoms in diazobenzene perbromide are more loosely combined

* *Ber.*, 1897, **33**, 2179. † *Ber.*, 1900, **33**, 2179.
‡ Euler, *Ber.*, 1895, **31**, 4168.

than the third; the compound was therefore regarded as having the constitution $C_6H_5 . N : NBr, Br_2$.

Kekulé looked upon this as a tribromohydrazine,

$$C_6H_5 . NBr.NBr_2,$$

and Erlenmeyer wrote it as

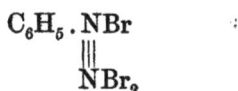

$$C_6H_5 . NBr$$
$$\underset{NBr_2}{\overset{|||}{}}$$

Hantzsch has prepared a large number of these perhalides,* thus—

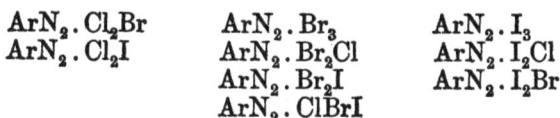

$ArN_2 . Cl_2Br$	$ArN_2 . Br_3$	$ArN_2 . I_3$
$ArN_2 . Cl_2I$	$ArN_2 . Br_2Cl$	$ArN_2 . I_2Cl$
	$ArN_2 . Br_2I$	$ArN_2 . I_2Br$
	$ArN_2 . ClBrI$	

and regards them as analogous to potassium tri-iodide, caesium tri-iodide, and the trihalides of the quaternary ammonium series. The exact arrangement of the three halogen atoms in the molecule is not known, and no isomerism such as $(Ar.N_2 . Cl + Br_2)$ and $(Ar.N_2Br + BrCl)$ exists.

§ 6. **Relation between** *syn-* **and** *anti-***compounds.**—The rate of isomerization of *syn-* to *anti-*diazo-compounds depends largely on the substituents present in the aromatic nucleus. Methyl groups hinder the rate, whilst halogen atoms increase it. Thus the transformation is very difficult to bring about in the case of trimethyl- and methoxy-benzenediazo-oxides, whilst in the case of the unsubstituted benzenediazo-oxide, $C_6H_5 . N_2 . OK$, it proceeds quickly above 100°. The *p*-bromo-derivative, on the other hand, is isomerized at boiling-point, the *p*-sulpho-derivative, $SO_3K . C_6H_5 . N_2 . OK$, slowly at the ordinary temperature, and the tribromo- and *p*-nitro-derivatives, $C_6H_2Br_3 . N_2 . OK$ and $NO_2 . C_6H_4 . N_2 . OK$, instantaneously, so that the *syn*-salt cannot be isolated.

In the case of the diazo-sulphonates it is the alkylated derivatives which isomerize more quickly than the parent substance, whilst the *p*- and *o*-halogen substituted derivatives of the *syn* series are relatively stable.*

* *Ber.,* 1895, **28**, 2754.
† Hantzsch and Schmiedel, *Ber.,* 1894, **27**, 3071, 3530.

Similarly among the *syn*-diazo-cyanides the *o*- and *p*-halogen substituted derivatives are fairly stable, and the parent sub-stance, $C_6H_5 . N_2 . CN$, has not been isolated.

The presence of the nitro-group greatly increases the rate of isomerization in all the above series of *syn*-compounds.

§ 7. The isomeric diazo-sulphonates and diazo-cyanides.— Having explained Hantzsch's views as to the constitution of the diazonium salts, we can now resume the discussion of the constitution of the diazo-sulphonates and diazo-cyanides which were described on pp. 128, 130.

The formulae given by Hantzsch to these compounds

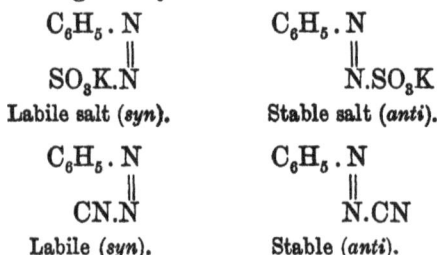

$$C_6H_5 . N \qquad\qquad C_6H_5 . N$$
$$\| \qquad\qquad\qquad \|$$
$$SO_3K.N \qquad\qquad N.SO_3K$$

Labile salt (*syn*).   Stable salt (*anti*).

$$C_6H_5 . N \qquad\qquad C_6H_5 . N$$
$$\| \qquad\qquad\qquad \|$$
$$CN.N \qquad\qquad N.CN$$

Labile (*syn*).   Stable (*anti*).

were objected to by Bamberger * and Blomstrand,† who did not accept the stereochemical hypothesis. These chemists assigned the following formulae to the above substances—

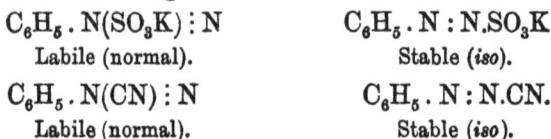

$$C_6H_5 . N(SO_3K) : N \qquad\qquad C_6H_5 . N : N.SO_3K$$

Labile (normal).   Stable (*iso*).

$$C_6H_5 . N(CN) : N \qquad\qquad C_6H_5 . N : N.CN.$$

Labile (normal).   Stable (*iso*).

the formulae for the two sulphonates having already been suggested by V. Meyer and Jacobson,‡ to which these authors have adhered in succeeding editions of their book.

Bamberger regarded the quinquevalency of nitrogen in the diazonium salts as being dependent on the negative character of the group with which the diazonium radical was united. Thus when these groups were Cl, NO_3, SO_4H, the nitrogen atom was necessarily quinquevalent. This condition was still maintained by the SO_3K and CN groups, but owing to their

* *Ber.*, 1895, **28**, 242, 447, 834.
† *J. pr. Chem.*, 1896 [ii], **53**, 169; 1897, **55**, 481.
‡ *Lehrbuch der org. Chem.*, II. 303.

small negative character the labile salts could pass into the stable salts (the nitrogen atom becoming tervalent) with extreme ease.

Hantzsch * reiterated his objection to this view of the constitution of the diazo-sulphonates from the fact that these salts are only dissociated into two ions, namely, $Ar.N_2.SO_3$ and K, whilst if they possessed the diazonium constitution they would, according to him, be expected to yield three ions, namely, $Ar.N_2$, SO_3, and K, corresponding to the behaviour of potassium sulphite, which gives the ions K, K, and SO_3.† Further, he pointed out that the colour of the normal diazo-sulphonate (red) was another argument against Bamberger's view, as benzenediazonium salts with colourless anions (as SO_3) were colourless.

A further argument against the stereochemical view of the isomerism of the diazo-sulphonates was adduced by von Pechmann.

He pointed out the fact that both the groups SO_3H and CN themselves could give rise to isomerism. In order to find a group free from this objection, von Pechmann ‡ selected the diazo-salts of benzenesulphinic acid, $C_6H_5.N{:}N.SO_2.C_6H_5$, which had been prepared by Koenigs.§

In whatever manner this salt was prepared, it was impossible to discover the existence of an isomeride; this was also true of the p-nitro-derivative, $NO_2.C_6H_4.N{:}N.SO_2.C_6H_5$, and von Pechmann concluded that these facts militated against Hantzsch's theory.

Hantzsch and Singer ‖ also prepared a number of these additive compounds, but were unable to detect the existence of isomerism. The supposed isomerism of the diazo-thiosulphonates ¶ was shown by Dybowski and Hantzsch ** to have no foundation in fact.††

* *Ber.*, 1895, **28**, 676.
† See however Ostwald's opinion, p. 130.
‡ *Ber.*, 1895, **28**, 861.   § *Ber.*, 1877, **10**, 1531.
‖ *Ber.*, 1897, **30**, 312.
¶ Tröger and Ewers, *J. pr. Chem.*, 1900 [ii], **62**, 369.
** *Ber.*, 1902, **35**, 268.
†† Compare also Hantzsch and Glogauer, *Ber.*, 1897, **30**, 2548; Hantzsch, *Ber.*, 1898, **31**, 636.

With regard to the constitution of the diazo-cyanides, Hantzsch * insisted that a diazonium cyanide, Ar.N(CN) : N, must be similar to an alkali cyanide, but as the normal diazo-cyanides were coloured, sparingly soluble in water, and soluble in organic solvents, they could not have the constitution attributed to them by Bamberger and Blomstrand.

Finally, Hantzsch and Danziger † succeeded in preparing a third series of cyanides by treating a diazonium chloride with a suspension of silver cyanide.   The insoluble yellow syn-diazo-cyanide is also formed in this reaction, but the filtrate contains a soluble double cyanide with silver cyanide, which is considered by Hantzsch to be a true diazonium derivative. These substances are soluble in water and colourless and resemble the alkali cyanides, and are therefore diazonium cyanides.

The formation of these double diazonium cyanides led Hantzsch to the hypothesis that the sparingly soluble syn-diazo-cyanides may exist in solution in a state of equilibrium with the isomeric diazonium salt, and a study of the diazo-cyanides derived from p-anisidine confirmed this idea.‡   p-Methoxy-benzenediazonium bromide and chloride with potassium cyanide in alcoholic solution yield the syn-diazo-cyanide

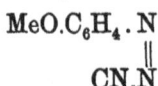

$$\text{MeO.C}_6\text{H}_4 . \text{N}$$
$$\|$$
$$\text{CN.N}$$

an orange-red, insoluble substance, melting at 51°, and coupling with β-naphthol.   This changes slowly into the anti-salt

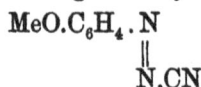

$$\text{MeO.C}_6\text{H}_4 . \text{N}$$
$$\|$$
$$\text{N.CN}$$

which is brownish red, melts at 121°, and does not couple with β-naphthol.

(Certain syn-cyanides are difficult to convert into the anti-modification.   Thus that derived from 2 : 4 : 6-tribromoaniline must be combined with benzenesulphinic acid to form the additive product $C_6H_2Br_3 . NH.N(CN).SO_2 . C_5H_5$, which on treatment with alkalis yields the anti-cyanide.)

* Ber., 1895, 28, 668.   † Ber., 1897, 30, 2529.
‡ Ber., 1900, 33, 2161 ; 1901, 34, 4166.

When, however, an aqueous solution of p-methoxybenzene-diazonium hydroxide is evaporated with excess of hydrogen cyanide at the ordinary temperature, a colourless crystalline substance is obtained which has the composition

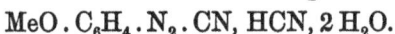

$$MeO . C_6H_4 . N_2 . CN, HCN, 2 H_2O.$$

This possesses all the properties of a true metallic salt, it is very soluble, and its solution is an electrolyte. It couples with β-naphthol, and is converted into the syn-diazo-cyanide by the action of alkaline solutions.

There can thus be prepared from p-anisidine three different diazo-cyanides, namely—

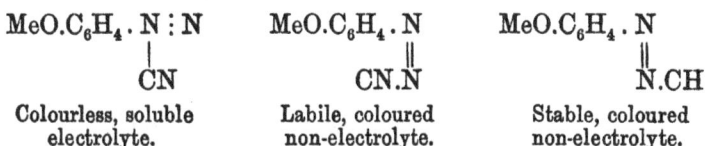

$MeO.C_6H_4 . N \vdots N$	$MeO.C_6H_4 . N$	$MeO.C_6H_4 . N$
$\mid$	$\parallel$	$\parallel$
CN	$CN.N$	$N.CH$
Colourless, soluble electrolyte.	Labile, coloured non-electrolyte.	Stable, coloured non-electrolyte.

The isolation of these three isomerides was regarded by Hantzsch as a very strong proof of his stereochemical theory, as Bamberger's theory could only account for two of them.[*]

It is, however, highly significant that in the two series of isomeric diazo-compounds, the cyanides and the sulphonates, both groups attached to the diazo-nucleus, should themselves be capable of giving rise to isomerism. As regards the cyanides, it has indeed been suggested [†] that Hantzsch's syn-compound has the constitution $Ar.N : N.NC$, and the $anti$-compound $Ar.N : N.CN$. (See also p. 141.)

This view would seem to be confirmed by the observation of Hantzsch and Schultze [‡] that both series give the same $(anti)$ diazobenzenecarboxylic acid, $Ar.N : N.CO_2H$, for the former compound would be expected to undergo transformation into the latter. According to the stereochemical theory the labile syn-compound would pass into the more stable $anti$-cyanide before hydrolysis. Moreover, Hantzsch has offered no proof against this obvious view; he contented himself with stating that neither of these compounds was an iso-cyanide.

[*] Another way of accounting for these is explained on p. 168.
[†] Orton, $Trans.$, 1903, 83, 805.   [‡] $Ber.$, 1895, 88, 2073.

§ 8. **Constitution of the metallic diazo-oxides.**—At the beginning of 1895 Bamberger's views on the constitution of diazo-compounds up to this time had been as follows:—The diazo-salts were to be represented by the Blomstrand formula—

$$C_6H_5 . NCl : N.$$

The normal, labile diazo-compounds (coupling with phenols) had the constitution $C_6H_5 . N : N.OX$.

The *iso*-diazo-compounds (nitrosoamines, not coupling with phenols) were $C_6H_5 . NH.NO$ or $C_6H_5 . NX.NO$, X being a metal such as K, Na, &c., but a little later he represented them as being divided into two groups, namely, (1) normal diazo-compounds (of diazonium type, $C_6H_5 . NCl : N$) ; (2) *iso*-diazo-compounds (of azo-type, $C_6H_5 . N : N.OH$).

Bamberger was led to this change of view by his work on the interaction of nitrosobenzene and hydroxylamine,* from which he supposed that the stable form of diazobenzene hydroxide was formed according to the equation

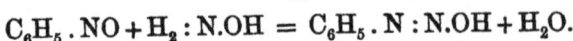

$$C_6H_5 . NO + H_2 : N.OH = C_6H_5 . N : N.OH + H_2O.$$

but, as we have shown (p. 133), in reality the normal or labile modification is produced.

The controversy existing in the years 1895 to 1897 between Hantzsch and Bamberger mainly resolved itself into a discussion of the constitution of the metallic diazo-salts. On the one hand Hantzsch strove to prove that they were stereoisomeric by means of physical measurements (electrical conductivity, &c.), whilst, on the other, Bamberger maintained that their differing chemical characteristics were sufficient evidence that they differed in constitution.

In 1895 Hantzsch and Gerilowski † prepared a labile form of the sodium salt of diazobenzenesulphonic acid,

$$NaO.N_2 . C_6H_4 . SO_3Na, 4H_2O,$$

the stable isomeride having been already obtained by Bamberger. This new labile form is obtained by treating the diazotized sulphanilic acid mixed with water with concentrated aqueous sodium hydroxide at 0°. It forms white, silky

---

* *Ber.*, 1895, **28**, 1218. † *Ber.*, 1895, **28**, 2002.

needles, has a strongly alkaline reaction, and couples instantly with β-naphthol. It becomes changed into the stable iso-meride (which contains no water and does not couple with β-naphthol) by heating with water. The labile salt in aqueous solution forms three ions, as does the stable salt; according to Hantzsch, if it were a diazonium compound it should form four ions.

In the following year Bamberger asserted that Hantzsch's conclusions as to the stereoisomerism of these two salts could not be maintained, as they were based on inaccurate observations of their behaviour.*

Further work was, however, done by Hantzsch. The deter-mination of the electrical conductivity of the two salts †️ showed that at moderate dilution ($v_{16}-v_{64}$) the conductivity was the same in each case. Whilst, however, the increase in the conductivity of the stable salt from v_{32} to v_{1024} corresponds with the theory for sodium salts of dibasic acids not hydroly-tically dissociated in aqueous solution,‡ a fact which shows that the stable salt is not hydrolysed, the conductivity of the labile salt from v_{128} increases very rapidly, thus showing that the labile salt has become hydrolysed, forming

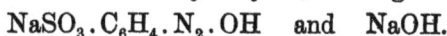

$$NaSO_3 . C_6H_4 . N_2 . OH \quad \text{and} \quad NaOH.$$

The solution also has an alkaline reaction, whilst that of the stable salt is neutral. The conclusion is, therefore, that both diazo-complexes possess acid properties, that of the labile salt being the weaker. The difference between the two salts is thus only a gradual one, and consequently Hantzsch con-sidered that they were stereoisomeric, assigning to them the constitution—

$$NaSO_3 . C_6H_4 . N \qquad NaSO_3 . C_6H_4 . N$$
$$\| \qquad\qquad \|$$
$$NaO.N \qquad\qquad N.ONa$$

Similarly, the cryoscopic researches of Goldschmidt § showed that both the normal (*syn*) and the *iso* (*anti*) potassium ben-zenediazo-oxides possessed the same number of ions in aqueous

* *Ber.*, 1896, **29**, 564. † *Ber.*, 1896, **29**, 743.
‡ *Zeitsch. physikal. Chem.*, 1894, **13**, 222.
§ *Ber.*, 1895, **28**, 2020.

L

solution, and this was likewise considered to be a proof of Hantzsch's view of their constitution.

Bamberger,[*] on the other hand, maintained his view that the two compounds were to be formulated—

$$C_6H_5 . N.OK$$
$$\text{|||}$$
$$N$$

Labile (normal).

$$C_6H_5 . N:N.OK$$

Stable (*iso*).

To Hantzsch's criticism that there existed no alkali metal the hydroxide of which possessed acid properties, Bamberger denied that diazonium was a compound alkali metal, and held that the hydroxide was neither comparable with tetramethylammonium hydroxide or with potassium hydroxide.

Bamberger's examples of chemical differences between the normal and the *iso*-salts, namely, that the *iso*-salt was reduced by sodium amalgam to phenylhydrazine and the normal not,[†] and that the *iso*-salt was converted into the normal salt when treated with benzoyl chloride, whilst the normal salt gave nitrosobenzanilide,[‡] were both shown by Hantzsch [§] to be based on error, as he obtained both phenylhydrazine and nitrosobenzanilide [||] in equal amounts in the two cases.

Later, however, Bamberger [¶] became convinced that a diazonium hydroxide could not act as an acid, and gave up the diazonium configuration for the normal metallic salts. He now regarded the metallic diazo-salts as existing in the two forms: (1) normal metallic diazo-salts (or diazotates), $Ar(N_2OK)$, of unknown constitution; the normal diazohydroxides, however, were, according to him,

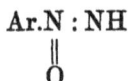   or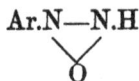

and (2) *iso*-diazotates, $Ar.N:N.OK$.

§ **9. Diazo-ethers.**—As has been explained (p. 98), von Pechmann and Frobenius [**] discovered that the sodium salt of

* *Ber.*, 1896, **29**, 457.
+ *Ber.*, 1896, **29**, 473.   ‡ *Ber.*, 1897, **30**, 211.
§ *Ber.*, 1897, **30**, 339.
|| *Ber.*, 1897, **30**, 621 ; 1899, **32**, 1718.
¶ *Annalen*, 1900, **313**, 97.   ** *Ber.*, 1894, **27**, 672.

p-nitrobenzenediazo-oxide (*iso*-compound of Schraube and Schmidt) gave with methyl iodide a nitrogen ether,

$$NO_2 . C_6H_4 . N(CH_3) . NO$$

but that the silver salt yielded an oxygen ether—

$$NO_2 . C_6H_4 . N:N.O.CH_3.$$

On this account they considered that nitro-*iso*-diazobenzene hydroxide was a tautomeric substance—

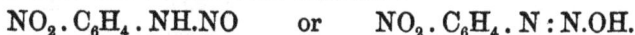

$$NO_2 . C_6H_4 . NH.NO \quad \text{or} \quad NO_2 . C_6H_4 . N:N.OH.$$

They regarded the oxygen ether, therefore, as a normal diazo-compound (although it was derived from the *iso*-salt), as it combined with phenols like diazo-salts, and the nitrogen ether as the *iso*-compound.

Hantzsch * regarded the oxygen ether as an *anti*-compound, but experiments by Bamberger † and von Pechmann and Frobenius ‡ confirmed the resemblance of this compound to the normal diazo-salts and its difference from the *anti*-compounds. Moreover, a large number of similar ethers were prepared by Bamberger,§ and these were also found to react as normal compounds; on hydrolysis with alkalis they yielded normal metallic derivatives. Shortly after, Bamberger ‖ considered that the diazo-ether ought to be regarded as an *iso*-compound, and Hantzsch and Wechsler ¶ found that *p*-bromo-diazobenzene ethyl ether yielded the *anti*-oxide on hydrolysis.

Some time later, as the conflicting views on this subject had not been entirely reconciled, Euler ** investigated the matter afresh. By careful experiment he found that the product of hydrolysis of diazobenzene methyl ether, as well as *p*-bromo-diazobenzene methyl ether, reacted, as did a normal diazo-compound, but Hantzsch was able to show that the coupling with α-naphthol, on which these experiments were based, was due to a secondary reaction, and that, in fact, the *iso* (*anti*) compounds were produced on hydrolysis. It appears, therefore, that von Pechmann's oxygen ether belongs to the *iso*- or *anti*-series.††

* *Ber.*, 1894, **27**, 1865, 2968.   † *Ber.*, 1894, **27**, 3412.
‡ *Ber.*, 1895, **28**, 170.   § *Ber.*, 1895, **28**, 225.
‖ *Ber.*, 1895, **28**, 829.   ¶ *Annalen*, 1902, **325**, 226.
** *Ber.*, 1903, **36**, 2503.
†† Hantzsch, *Ber.*, 1903, **36**, 3097, 4361; 1904, **37**, 3030; Euler, *Ber.*, 1903, **36**, 3835.

§ 10. **Diazo-anhydrides.**—In 1896 Bamberger * discovered that when normal metallic diazo-salts are treated with cold dilute acetic acid, extremely explosive, yellow diazo-anhydrides are formed. These cannot be obtained from the *iso*-salts, which yield colourless hydroxides under similar conditions, and this difference was considered by Bamberger to be another proof of the structural difference of the two.

The diazo-anhydrides may also be prepared in some cases by treating a diazonium salt with a normal metallic diazo-salt. They couple slowly with phenols, yield oxygen ethers with the alcohols, and react explosively with benzene, yielding diphenyl derivatives. With alkalis they yield the corresponding normal salt, and mineral acids convert them into diazonium salts.

With amines, diazoamino-compounds are obtained; with ammonia, bisdiazoamino-compounds; and with bromine, diazo-perbromides.

Bamberger was of the opinion that their constitution was to be represented by

$$\text{R.N.O.N.R}$$
$$\overset{\|}{\text{N}} \quad \overset{\|}{\text{N}}$$

but Hantzsch † considered that they were more probably represented by R.N : N.O.N : N.R. He found later ‡ that the diazo-anhydrides readily yield *syn*-diazo-cyanides on treatment with hydrogen cyanide, and pointed out that the anhydrides dissolve very slowly in hydrochloric acid to form diazonium chlorides—facts which confirmed his theory of their azo-constitution.

Bamberger later § suggested the formula

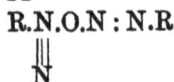

$$\text{R.N.O.N : N.R}$$
$$\overset{\|}{\text{N}}$$

but this was rejected by Hantzsch on the ground that, the *syn*-diazo-hydroxide being an extremely weak acid, such a

* *Ber.*, 1896, **29**, 446.
† *Ber.*, 1896, **29**, 1074; 1897, **30**, 626.
‡ *Ber.*, 1898, **31**, 636.   § *Ber.*, 1898, **31**, 2636.

diazonium diazo-oxide should be instantly decomposed by acids.

From the fact that the diazo-anhydrides yielded *syn*-diazo-cyanides with hydrogen cyanide and *syn*-diazo-sulphonates with potassium sulphite, Hantzsch adopted the *syn*-formula—

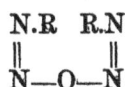

$$
\begin{array}{ccc}
\text{N.R} & & \text{R.N} \\
\| & & \| \\
\text{N} & \!\!\!\!-\text{O}-\!\!\!\! & \text{N}
\end{array}
$$

§ 11. **Diazo-hydroxides.**—Up to the year 1898, although the existence of isomeric metallic diazo-oxides was without doubt, the free diazo-hydroxides corresponding to these had not been prepared.

From the great similarity of the diazonium salts to the ammonium salts, Hantzsch drew the conclusion that a corresponding diazonium hydroxide should be capable of existence, which would of course make a third isomeric hydroxide, having the constitution $C_6H_5 . N(OH) : N$. He succeeded in obtaining an aqueous solution of this by treating diazobenzene chloride with silver oxide (see p. 100). Determination of the electric conductivity of the solution * showed that the affinity constant of the base at 0° is seventy times greater than that of ammonium hydroxide, and is a little greater than that of piperidine. The affinity constants of methoxybenzenediazonium hydroxide and ψ-cumenediazonium hydroxide are even greater, and are very close to those of the alkali hydroxides.

The effect of introducing halogens into the aromatic nucleus is shown in the following table:—

	k = velocity constant.
$C_6H_5 . N_2 . OH$	0·123
$Br.C_6H_4 . N_2 . OH$	0·0149
$(2:4) Br_2 : C_6H_3 . N_2 . OH$ . . .	0·0136
$(2:4:6) Br_3 : C_6H_2 . N_2 . OH$ . .	0·0014

A comparison of the electrical conductivity experiments with the results obtained in the hydrolysis of ethyl acetate

* Hantzsch and Davidson, *Ber.*, 1896, **31**, 1612.

by benzenediazonium hydroxide indicates that, in $1/128$ N-solution at $0°$, about 33 per cent. of the base exists in the ionized condition. The ionization observed in the hydrolysis experiments is greater than that determined by the conductivity experiments, and this shows that the electrolytic dissociation is entirely due to the reaction

$$C_6H_5.N_2.OH \; \rightleftarrows \; C_6H_5.N:N+OH$$

and not to the electrolysis of a diazonium *syn*-diazo-oxide—

$$C_6H_5.\underset{\overset{|||}{N}}{N}.O.N_2.C_6H_5$$

§ 12. Condition of the non-ionised diazonium hydroxide.—

The solution of benzenediazonium hydroxide, when treated with alkali hydroxides, generates an appreciable amount of heat, and thus behaves as a weak acid. This reaction is also indicated by determinations of the electrical conductivity of the diazonium hydroxide solutions when treated with one, two, or more molecular proportions of sodium hydroxide.

Hantzsch and Davidson explain this by assuming that the non-ionized part of the diazonium hydroxide exists in solution in a hydrated form, thus—

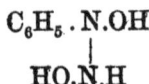

$$C_6H_5.N.OH$$
$$|$$
$$HO.N.H$$

which, with alkali hydroxide, loses water, giving the *syn*-diazo-hydroxide

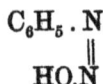

$$C_6H_5.\underset{\overset{||}{HO.N}}{N}$$

and this then furnishes the sodium salt—

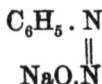

$$C_6H_5.\underset{\overset{||}{NaO.N}}{N}$$

Diazonium hydroxides are consequently known only in solution, and the existence of *syn*-diazo-hydroxides is doubtful.

§ 13. Constitution of *iso* (*anti*) diazo-hydroxides.—In

1899 Hantzsch enunciated his theory of pseudo-acids, a term

applied to neutral compounds which, under the influence of alkalis, yield stable salts. Thus, for example, phenylnitromethane, $C_6H_5.CH_2.NO_2$, is stable, neutral, and a non-electrolyte, but with alkalis it changes to the isomeric form $C_6H_5.CH:NO.OH$, which forms stable salts, thus—

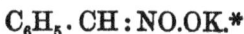

$$C_6H_5.CH:NO.OK.*$$

An examination of the properties of the metallic *anti*-diazo-oxides showed that the solution obtained by treating them with an equivalent amount of hydrochloric acid has a neutral reaction, and, conversely, when this solution is treated with an equivalent quantity of alkali, the product is neutral. The substance obtained, therefore, by treating the diazo-salt with acid has the properties of a pseudo-acid,† and is best represented as being a primary nitrosoamine—

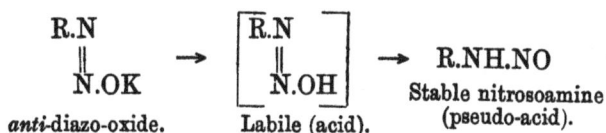

$$\begin{array}{c} R.N \\ \parallel \\ N.OK \end{array} \rightarrow \left[\begin{array}{c} R.N \\ \parallel \\ N.OH \end{array}\right] \rightarrow \begin{array}{c} R.NH.NO \end{array}$$

| *anti*-diazo-oxide. | Labile (acid). | Stable nitrosoamine (pseudo-acid). |

In this, it will be noticed, Hantzsch adopts the older view, so long combated by him, of the tautomeric form of the *iso*-diazo-hydroxide, with the exception that he adheres to the *anti*-configuration for the labile form.

Hantzsch and Pohl ‡ claimed to have prepared these *anti*-diazo-hydroxides and also the nitrosoamines in the solid condition, and stated that 'the nature of the so-called *iso*-diazo-hydroxides is now definitely elucidated'. This work, however, cannot be regarded as having any bearing on the point, as one example dealt with by Hantzsch and Pohl, namely, the conversion of the metallic *anti*-salt of 2 : 4 : 6-tri-bromodiazobenzene into the nitrosoamine, was shown by Orton § to be quite inaccurate. Orton found that the substance described as the nitrosoamine by Hantzsch and Pohl was in reality a mixture of the quinonediazide,

---

* *Ber.*, 1899, **32**, 575.
† Hantzsch, Schumann, and Engler, *Ber.*, 1899, **32**, 1703.
‡ *Ber.*, 1902, **35**, 2964.
§ *Proc. Roy. Soc.*, 1902, **71**, 153 ; *Trans.*, 1903, **83**, 796.

$$\text{Br} \overset{\overset{\displaystyle \text{N}_2}{\diagup\diagdown}}{\underset{\underset{\displaystyle \text{Br}}{\bigvee}}{\ }} :\text{O}$$

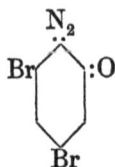

(see p. 67), and a hydroxyazo-compound.*

This proof, of course, must be held to throw grave doubt on the correctness of the other cases mentioned by Hantzsch and Pohl, especially as Hantzsch has admitted that 2 : 4 : 6-tri-bromophenylnitrosoamine is unstable and cannot be isolated free from other substances in an analysable condition. Orton's work thus shows that no nitrosoamine is formed under the conditions used by Hantzsch.

* Compare also Hantzsch, *Ber.*, 1903, **36**, 2069; Orton, *Trans.*, 1905, **87**, 99.

CHAPTER XIX

OTHER VIEWS OF THE CONSTITUTION OF THE DIAZO-COMPOUNDS FROM 1895

§ 1. **Constitution of the coloured diazo-salts of Jacobson.**
—In 1895 Jacobson * examined the diazo-salts of p-amino-diphenylamine which had been first prepared by Ikuta.† These diazo-salts are distinguished by their great stability and by their yellow colour, in consequence of which Jacobson assigned to them the constitution—

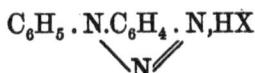

$$C_6H_5 . N.C_6H_4 . N,H\overset{\cdot\cdot}{X}$$
$$\diagdown N \diagup$$

Hantzsch investigated the reactions of these compounds, pointing out that other coloured diazo-salts were known which had undoubtedly the normal constitution, namely, the diazo-salts of di-iodobenzene, diazofluoren, diazophenanthrene, &c.

He showed that Jacobson's diazo-salts had a neutral reaction and were thus similar to the ordinary diazo-salts, whilst a compound of the above formula would be expected to undergo hydrolytic dissociation and therefore show an acid reaction. The salts were therefore considered by Hantzsch to possess the constitution $C_6H_5 . NH.C_6H_4 . NX : N$.

By the action of potassium hydroxide, however, no corresponding metallic salt was formed, but an explosive compound insoluble in water, having the formula $C_{12}H_9N_3$, which was evidently an anhydride of the diazo-hydroxide,

$$C_6H_5 . NH.C_6H_4 . N_2 . OH$$

and to which Hantzsch gave the formula

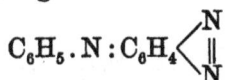

$$C_6H_5 . N : C_6H_4 \diagup \!\!\! \overset{N}{\underset{N}{\big\|}}$$

* *Annalen,* 1895, **287**, 131.   † *Annalen,* 1893, **272**, 282.

corresponding to Wolff's formula * for the quinonediazides—

$$O : C_6H_4 \diagdown \begin{matrix} N \\ \| \\ N \end{matrix}$$

§ 2. Constitution of diazo-salts according to Walther.—

A formula for diazobenzene chloride was proposed by Walther in 1895,† but has not hitherto found acceptance.

Walther, in endeavouring to explain the fact that the same product is formed by the interaction of diazobenzene chloride and bromoaniline, on the one hand, and bromodiazobenzene chloride and aniline, on the other, suggested that nitrous acid might be supposed to contain quinquevalent nitrogen, and represented the formation of diazobenzene chloride by the equation—

$$C_6H_5 . NH_3Cl + N \diagup\limits^{O}\diagdown_{H} O = C_6H_5 . NH_2Cl.N \diagup\limits^{O}\diagdown_{H} OH$$

$$= C_6H_5 . NHCl : NH : O + H_2O$$

this representing diazobenzene chloride only in aqueous solution. The hydroxide would hence be $C_6H_5 . N : NH : O$, and diazoaminobenzene $C_6H_5 . N : NH : N . C_6H_5$, thus providing an explanation of the fact referred to above.

§ 3. Constitution of diazo-compounds according to Brühl.

—The question of the constitution of the diazo-compounds has been attacked by Brühl from the point of view of their refractive powers.‡

It was found that the refraction of the N_2 group in the diazo-compounds is 8·41, or about 3·4 higher than that of the same group in the primary hydrazines.

In azoxybenzene, the value for the N_2O group is 11·9, whilst that calculated on the assumption of a single linking between the nitrogen atoms is 7·5, so that Brühl regards azoxybenzene as a compound of the structure—

$$O \diagdown \begin{matrix} N.C_6H_5 \\ \| \\ N.C_6H_5 \end{matrix} \quad \text{or} \quad \begin{matrix} C_6H_5 . N : N.C_6H_5 \\ \| \\ O \end{matrix}$$

---

* *Annalen*, 1900, **312**, 126.
† *J. pr. Chem.*, 1895 [ii], **51**, 528, 581.
‡ *Zeitsch. physikal. Chem.*, 1898, **25**, 577, **26**, 47.

The normal diazo-oxides are looked upon as being constituted similarly to the nitrosoacyl-compounds, and the formation of diazobenzene from nitrosoacetanilide is written—

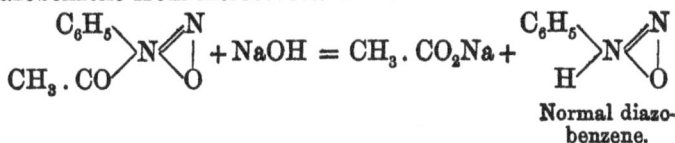

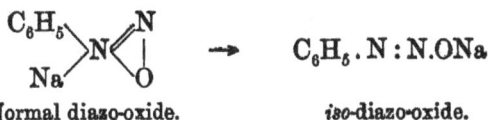

$$C_6H_5 >N-N< \begin{matrix} | \\ O \end{matrix} + NaOH = CH_3.CO_2Na + \quad C_6H_5 >N-N< \begin{matrix} | \\ O \end{matrix}$$
CH₃.CO   H

Normal diazo-
benzene.

$$C_6H_5 >N-N< \begin{matrix} | \\ O \end{matrix} \quad \rightarrow \quad C_6H_5.N:N.ONa$$
Na

Normal diazo-oxide.   *iso*-diazo-oxide.

Diazonium salts, however, have the Blomstrand formula $C_6H_5.NCl:N$. *p*-Nitrodiazobenzene methyl ether (compare p. 147) has the constitution $NO_2.C_6H_4.N:N.OMe$ and is an *iso*-compound.

Finally, benzenediazoic acid is regarded as possessing the nitroamine constitution—

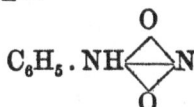

$$C_6H_5.NH \begin{matrix} O \\ < \diamond > N \\ O \end{matrix}$$

§ 4. Constitution of the diazo-compounds according to Dobbie and Tinkler.—Dobbie and Tinkler * attacked the problem of deciding the constitution of the isomeric diazo-compounds by observing their ultraviolet absorption spectra.

The two forms of benzaldoxime had previously been shown to exhibit identical spectra, so that isomeric substances, differing only as do the benzaldoximes, should also give identical spectra, but distinct ones if they were structurally isomeric.

Diazo-sulphonates.—The potassium benzenediazo-sulphonates were found to give identical spectra, which would be expected if the substances had the constitution assigned to them by Hantzsch, namely—

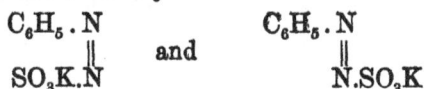

$$\begin{matrix} C_6H_5.N \\ \| \\ SO_3K.N \end{matrix} \quad and \quad \begin{matrix} C_6H_5.N \\ \| \\ N.SO_3K \end{matrix}$$

Diazo-cyanides.—The diazo-cyanides prepared from *p*-anisidine and *p*-chloroaniline were examined; both of these pairs

* *Trans.*, 1905, 87, 273.

of isomerides gave almost identical spectra, so that here again the *syn-* and *anti*-configuration would account for this.

The solution of the diazonium cyanide, $OMe.C_6H_4.N(CN):N$, gave an entirely different spectrum. This compound is therefore structurally isomeric with the other two.

Diazo-oxides.— The potassium benzenediazo-oxides were found to give quite different spectra, and the conclusion is that they are structurally isomeric. The nitrosoamine formula for the more stable salt would account for this difference, and it was found that the spectrum of this salt and that of phenylmethyl-nitrosoamine, $C_6H_5.N(CH_3).NO$, were in complete agreement, a fact which points to the formula $C_6H_5.NK.NO$ as the correct one for the stable salt.

It was further discovered that a very dilute solution of the labile compound had a spectrum agreeing closely with that of diazobenzene chloride, and this appears to indicate that the original compound changes into a third modification, having the constitution of a true diazonium compound—

$$C_6H_5.N(OK):N.$$

Sulphobenzenediazo-oxides.— The potassium and sodium compounds obtained from diazotized sulphanilic acid by the action of caustic alkali were also examined. These gave similar results, as in the preceding case; the spectra were different and they are therefore structurally isomeric and not stereoisomeric.

Applying the reasoning used by Dobbie and Tinkler in the preceding case, these compounds would consequently possess the constitution—

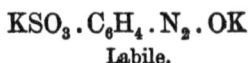

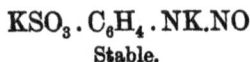

$$KSO_3.C_6H_4.N_2.OK \qquad KSO_3.C_6H_4.NK.NO$$
Labile.   Stable.

§ 5. Constitution of the diazo-compounds according to Armstrong and Robertson. — Armstrong and Robertson, in 1905,[*] in discussing the question of the relation of colour to constitution, considered that the yellow colour of phenylazo-ethane, $C_6H_5.N:N.C_2H_5$, is conditioned by the presence of the group $C_6H_5.N:N.$ alone, the ethyl radical not being known

* *Trans.*, 1905, **87**, 1280.

as a chromogenic centre in any other case. Arguing from this, they concluded that all compounds of the form

$$C_6H_5 . N : N.X$$

should be coloured, and consequently that only coloured diazo-compounds can be represented by such a formula. Armstrong and Robertson call the above compound 'phenyldiazoethane', but it belongs to the azo-group just as much as does azobenzene, $C_6H_5 . N : N.C_6H_5$. Moreover, there are great chemical differences between the coloured diazo-compounds

$$C_6H_5 . N_2 . X$$

and the azo-compounds, $C_6H_5 . N_2 . R$, where X is an acidic group and R is an inert group, and, to take an example, according to the above reasoning, if coloured diazo-compounds of the formula $C_6H_5 . N : N.X$ unite readily with phenols, &c., to form azo-compounds, one should expect all compounds containing the group $C_6H_5 . N : N.$ to give the same reaction, which of course they do not. It is therefore not correct to compare the two in the way Armstrong and Robertson have done. These authors, from the above reasoning, deny that the *syn*- and *anti*-formulae can represent the constitutions of the labile and stable metallic diazo-compounds respectively, and adopt the nitrosoamine formula R.NK.NO for the latter.

For the colourless diazo-salts the diazonium formula is advocated as being the only alternative one which, at the time, could be devised. The isomeric sulphonates and cyanides are considered to be represented by the formulae—

R.N.SO$_3$K
‖
N
Labile.

R.N : N.SO$_3$K
Stable.

and

R.N.CN
‖
N
Labile.

R.N : N.CN
Stable.

In the case of the cyanides, Hantzsch's *syn*-compound is regarded as a mixture of the diazonium salt and the *anti*-salt.

In a similar way the labile metallic compounds are assigned the diazonium configuration, whilst the stable compounds, which, being colourless, could not be written R.N:N.OH according to the authors' reasoning, are considered to be further hydrated, R.N(OH).NH(OH) or R.NH.N(OH)$_2$, and these diazo-hydrates, on dehydration, would give rise to the isodynamic nitrosoamines, thus—

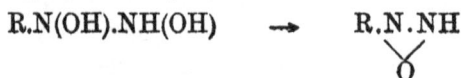

$$R.N(OH).NH(OH) \longrightarrow R.N.NH$$
$$\diagdown O \diagup$$

$$R.NH.N(OH)_2 \longrightarrow R.NH.NO.$$

Which of these is the parent substance of the *iso*-compounds is considered to depend on the colour or non-colour of pure nitrosoamines.

These views provoked a vigorous criticism by Hantzsch,[*] who pointed out that both coloured and colourless azo-compounds exist in the aliphatic series, for example, the deep red azo-dicarboxylic ester, $CO_2R.N:N.CO_2R$, and the colourless azo-*iso*-butyric acid derivatives, $CRMe_2.N:N.CRMe_2$,[†] and in the aromatic series, the nitrodiazo-ester

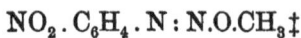

$$NO_2.C_6H_4.N:N.O.CH_3[‡]$$

is quite colourless.

He maintained, therefore, that the presence of the group .N:N. was no reason why a compound should be coloured.

The diazonium formula for the normal diazo-oxides, and Armstrong and Robertson's proposed formulae for the *iso*-diazo-oxides had previously been shown to be unsatisfactory and, as the *syn*-diazo-cyanides and sulphonates are more intensely coloured than the *anti*-forms, the former could not consist of a mixture of the latter with a colourless diazonium salt.

---

* *Proc.*, 1905, **21**, 289.   † Thiele, *Annalen*, 1896, **290**, 1.
‡ von Pechmann, *Ber.*, 1894, **27**, 672.

CHAPTER XX

A REVIEW OF THE VARIOUS THEORIES OF THE DIAZO-COMPOUNDS TO 1907

IT is evidently an impossible task to reconcile all the conflicting theories of the constitution of diazo-compounds, and although some of them may be dismissed at once, others must receive careful consideration.

In spite of the immense amount of work done in this field of research by Hantzsch and his pupils, it cannot be said that the stereochemical theory is generally accepted—there are many evidences in chemical literature which point to this conclusion. On the other hand, the alternative view of structural isomerism has several exponents, but no common ground has apparently been reached.

We shall, therefore, endeavour to sum up the principal points connected with the diazo-compounds, which are important in arriving at a theory of the constitution of these compounds.

§ 1. **Constitution of the diazo-salts (diazonium salts).**— There is fairly general agreement that the formula of Blomstrand represents the constitution of the diazo-salts better than that proposed by Kékulé. The existence of a salt-forming nitrogen atom in the diazo-complex makes it necessary to assume that one nitrogen at least is quinquevalent. Moreover, it would appear most probable that this nitrogen atom is the one attached directly to the aromatic nucleus, for otherwise we should arrive at the formula $C_6H_5 . N : NCl$, which, postulating as it does a quadruple linking between the two nitrogen atoms, is unlikely. We thus arrive at the conclusion that in diazobenzene chloride there is a union between the phenyl group and a quinquevalent nitrogen atom, which is linked to a univalent chlorine atom and united with

a second nitrogen atom. This union may be indicated by
a dotted line thus—

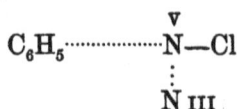

$$C_6H_5 \cdots\cdots\overset{\text{v}}{N}-Cl$$
$$\overset{\vdots}{N}\text{ iii}$$

It is very important to observe, however, that the facts
referred to do not prove anything more than this; that is to
say, they do not indicate the *number* of bonds between phenyl
and quinquevalent nitrogen or between quinquevalent and
tervalent nitrogen. It has, however, been generally considered
that the union between phenyl and quinquevalent nitrogen is
one linking, whilst that between the two nitrogen atoms is
three, thus—

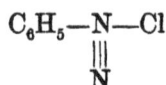

$$C_6H_5-\overset{|||}{N}-Cl$$
$$N$$

There is, however, another and, in the opinion of the
author, a better way of arranging these linkings (see p. 163).

§ 2. **The labile and stable isomeric diazo-compounds.**—In
studying the properties of these compounds it is impossible
to avoid the conclusion that the labile or normal compounds
resemble the diazonium salts most closely. The similar
behaviour with regard to the formation of azo-compounds,
the instability, the probability that, in many cases, equilibrium
mixtures or solid solutions of the two exist, and the resemblance of the two absorption spectra all point to this
direction.

The constitution of the labile salts should therefore be more
closely allied to that of the diazonium salts than to that of
the stable salts. Probably for this reason V. Meyer and
Jacobson, Blomstrand and others, regarded the labile compounds actually as diazonium compounds.

It seems, however, probable that Hantzsch's reasoning
against this view as regards the metallic compounds (p. 135)
is correct, that is to say, they are not diazonium derivatives.

In this connexion the views of Brühl as to the close
relationship between the constitution of the normal metallic

diazo-compounds and the nitrosoacyl-compounds are to be noted, and it seems likely that the true constitution of the former may be similar to that of the latter, although not assuming the form assigned to them by Brühl, and that the formulae of the nitrosoacyl-compounds may be tautomeric with those of the corresponding diazo-acetates. Such a new formulation of the nitrosoacyl-compounds, however, cannot be suggested up to the present.

We arrive, therefore, at the conclusion that neither the actual diazonium formula $C_6H_5 . N(OK):N$ nor the *syn*-diazo-formula

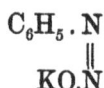

$$C_6H_5 . N$$
$$\overset{\|}{KO.N}$$

really represents the constitutions of these compounds.

The labile sulphonates are also probably not diazonium compounds, but the arguments against the possibility of them being sulphites are not so strong, and this is one of the reasons why the stereochemical theory is not accepted with regard to this case.

Hantzsch's argument against the sulphite constitution is principally that a diazo-sulphite would form three ions whilst the sulphonates give rise to only two ions. This argument has been recorded in many textbooks without, however, the important fact being added that no less an authority than Ostwald has stated (p. 130) that a diazonium sulphite would give rise only to two ions. Here, therefore, the possibility of the normal sulphonates being really sulphites cannot be regarded as excluded.

In the case of the labile diazo-cyanides we are met with a similar uncertainty as in the case of the sulphonates, namely, the possibility of isomeric change in the added group. So long as it is not proved that the labile diazo-cyanides cannot be isocyanides it cannot be maintained that they are *syn*-compounds.

Turning now to the stable or *iso*-compounds, most of the work done points to the nitrosoamine formula for the metallic compounds, $C_6H_5 . NK.NO$, and there seems to be a consensus

M

of opinion that the *iso*-diazo-hydroxides, diazo-cyanides, and sulphonates have the azo-constitution—

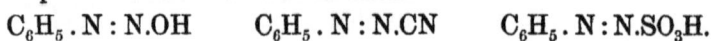

$$C_6H_5 . N : N.OH \qquad C_6H_5 . N : N.CN \qquad C_6H_5 . N : N.SO_3H.$$

From what has been said previously it will be evident that the existence of a special *anti*-configuration of these compounds depends on the simultaneous existence of the corresponding *syn*-compounds, which, as has been shown, cannot be regarded as having been definitely proved to possess this constitution.

APPENDIX

A NEW THEORY OF THE CONSTITUTION OF THE DIAZO-COMPOUNDS

In 1907 the author of this book put forward a new theory of the constitution of diazo-compounds,* which it is considered will not only explain the reactions of the diazo-compounds more readily than any of its predecessors, but also serve to throw light on some phenomena hitherto left unsolved. Perhaps the most striking reaction of the diazo-salts (diazonium salts) is the readiness with which the whole of the diazo-nitrogen is eliminated. There are no examples in the literature of singly-linked nitrogen being otherwise than firmly attached to the benzene nucleus and requiring energetic treatment for its liberation.

There are, however, cases where nitrogen, when attached to the benzene nucleus by two bonds, is most readily eliminated, one of the most striking being that of quinonechloroimide—

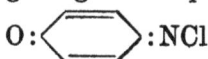

$$O:\langle \underline{} \rangle:NCl$$

Here, as in the case of the diazo-salts, the nitrogen is split off simply by heating the compound with water to 100°.

This reaction showing the great difference in behaviour of nitrogen attached by one and two linkings respectively to the aromatic nucleus, is obviously of much importance in arriving at a decision as to the manner in which nitrogen is united with the aromatic nucleus in diazo-salts.

It appears almost certain, from this analogy, that an atom of nitrogen in these salts is attached to the aromatic nucleus by two linkings. This idea at once leads us to a quinonoid configuration of the diazo-salts, thus—

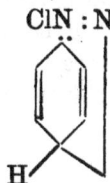

* *Trans.*, 1907, 91, 1049.

which conform to the requirements of these salts in that the
nitrogen attached to the benzene ring is quinquevalent,
and, of course, it explains more satisfactorily than does the
Blomstrand formula the ready elimination of diazo-nitrogen.

An obvious criticism, and indeed one which has been privately
advanced against this formula, is that, with a single linking
between the tervalent nitrogen atom and the para-carbon
atom, one should expect that, on reduction, the double linking
between the nitrogen atom would break and a para-diamine
result. This objection would be a weighty one were the
tervalent nitrogen united to a carbon atom simply (as in the
case of aniline) instead of to the CH group. This fact is of
much importance, for that a great difference in stability exists
in the two cases has been proved by E. Buchner. In his
researches on the action of diazoacetic ester on unsaturated
acid esters * this chemist found that the group

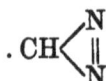

$$. \text{CH} \Big\langle \begin{matrix} \text{N} \\ \| \\ \text{N} \end{matrix}$$

always became converted into the group : C : N.NH, and it
may be concluded therefore that the latter group is more stable
than the former.

But we have in the new formula for diazo-salts (I) the
same group (II)

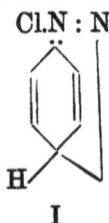

$$\text{Cl.N : N}$$

:C:N:N.CH.

I   II

in which, from the above work by Buchner, we may reason-
ably conclude that the linking : N.CH is more unstable than
the linking : C : N, and would be the first to be ruptured in
any reaction tending to destroy the configuration.

The first stage, therefore, in such a reaction can be represented by

* *Ber.*, 1894, **27**, 868, 877, 879 ; see also Curtius, *Ber.*, 1896, **29**, 767.

$$\text{ClN : N}$$

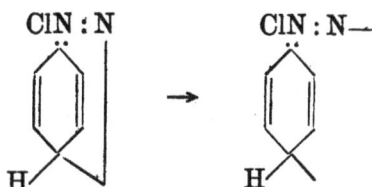

and the quinonoid formation having been thus disturbed, the ordinary configuration is resumed when the reaction proceeds to the next stage (reduction, formation of azo-compounds, elimination of nitrogen, &c.).

The ordinary reactions of the diazo-salts are thus satisfactorily explained.

All the work of Hantzsch and his collaborators on the nature of the radical 'diazonium', as has been shown, indicates that this acts like a compound alkali metal, the acid radical attaching itself to the quinquevalent nitrogen atom.

It is quite obvious that in this respect no difference can be detected between

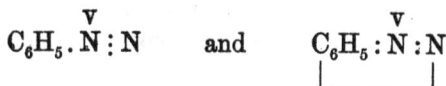

$$\overset{\text{v}}{C_6H_5}. N : N \qquad \text{and} \qquad \overset{\text{v}}{C_6H_5} : N : N$$

so that Hantzsch's results are equally applicable to the new formulation of 'diazonium'.

We shall now consider some phenomena in diazo-chemistry which have hitherto remained unexplained by any of the former theories. When we compare p-phenylenediamine and benzidine

we find a great difference in their behaviour towards nitrous acid. The former is converted into the tetrazo-compound only with difficulty and under special conditions, but the latter changes with perfect readiness. According to the Blomstrand or Kekulé formula this cannot be explained, but light is thrown on the mechanism of the reaction by the following considerations. Benzidine, when tetrazotized, becomes

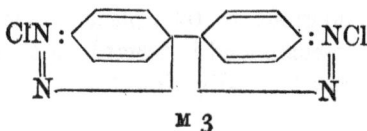

M 3

but the first stage in diazotizing p-phenylenediamine must give a compound of formula—

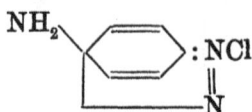

We now obtain a compound containing an amino-group, which is in the para-position with respect to a carbon atom, all of whose affinities are satisfied, and therefore it cannot link up with a second nitrogen atom. This explains why the tetrazotization does not proceed normally. Under the special conditions necessary, however (see p. 21), the linking between the aminic carbon atom and the tervalent diazo-nitrogen atom is broken, thus—

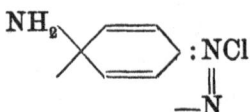

and now the amino-group can be diazotized, for its added nitrogen atom unites with the corresponding nitrogen atom of the first diazo-group—

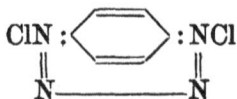

There are also several p-diamines in which only one amino-group can be diazotized, thus—

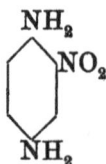

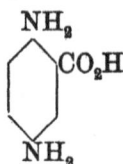

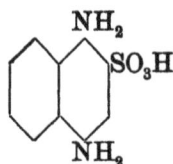

Here the same explanation as that given for the case of p-phenylenediamine probably holds good, but the para-linking may be rendered more stable by the presence of the acidic groups, and hence it does not break to allow the diazotization of the second amino-group to take place. This is, however, broken when an azo-compound is formed, so that the second amino-group may now be readily diazotized.

It is evident that, according to this theory, diazo-salts cannot be formed where a quinonoid configuration is precluded, so that we can now explain why the compounds

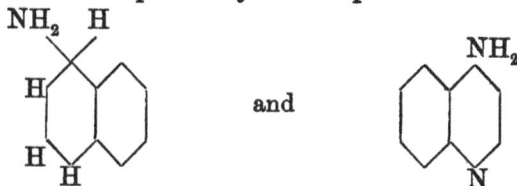

and

do not give diazo-salts, whilst the compounds

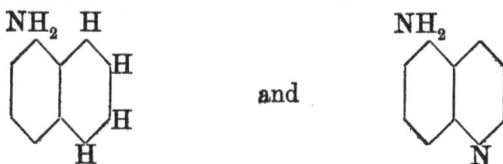

and

are readily diazotized.

In this connexion, also, we see that it is impossible for aliphatic amines to yield diazo-salts, thus

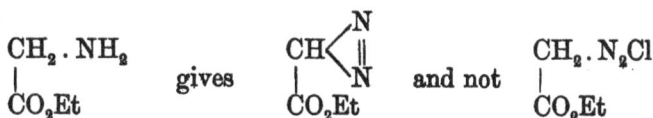

$$CH_2 \cdot NH_2 \quad \text{gives} \quad CH{<}\begin{matrix}N\\\|\\N\end{matrix} \quad \text{and not} \quad CH_2 \cdot N_2Cl$$
$$\underset{CO_2Et}{|} \qquad\qquad \underset{CO_2Et}{|} \qquad\qquad\qquad \underset{CO_2Et}{|}$$

This new formula for diazo-salts has received confirmation by the work of Morgan and Hird.*

Isomeric diazo-compounds.—The hydroxide corresponding with diazobenzene chloride would, on the above formulation, have the constitution—

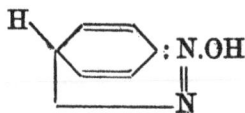

It is evident that the hydroxyl group may migrate to the other nitrogen atom now that the quinquevalency of the first nitrogen atom is not supported by the presence of an acidic group. We thus arrive at the formula

* *Trans.*, 1907, **91**, 1505 ; compare also Morgan and Wootton, *Trans.*, 1907, **91**, 1811.

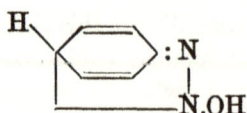

which, from its close connexion with the previous one, is an
exceedingly probable one for the normal (*syn*) diazo-compounds
(metallic salts, and, supposing that the normal cyanides and
sulphonates are not isocyanides and sulphites respectively,
for these also).

The great resemblance existing between the normal diazo-
compounds and the diazonium salts is very readily explained
by this formula. The more energetic means necessary to
produce the *iso*-diazo-compounds naturally tend to destroy the
bicyclic system here shown ; accordingly the change from
normal to *iso*-compounds occurs thus

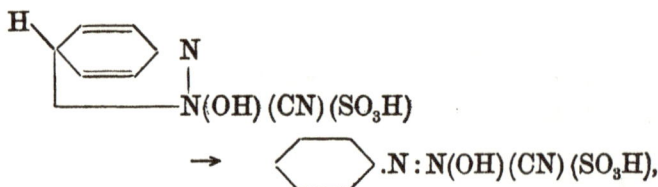

arriving in a very natural manner at the most probable formula
for the *iso*-compounds.

As has been shown, the formula for the *iso*-metallic salts
can allow tautomerism to take place, and consequently the
most stable condition is assumed by these compounds—

$$C_6H_5 . N : N.OK \quad \longrightarrow \quad C_6H_5 . NK.NO.$$

Finally, as the stereochemical theory appears thus to be
rendered unnecessary, it seems best to use the older terms,
normal and *iso*, instead of *syn* and *anti* respectively.

For the diazo-salts with acids, it is, however, most convenient
to retain the term ' diazonium', although in this book, as was
explained in the introduction, the term ' diazo ' introduced by
Griess, and also used by the Chemical Society, has been
retained in order to avoid confusion or misunderstanding before
the theoretical explanation had been reached.

SUBJECT INDEX

Absorption spectra of diazo-compounds, 155.
Acetoxy-group, replacement of diazo-group by, 53.
Alcohols, action of, on diazo-compounds, 38.
Amines, diazotization of, 14.
Aminoazo-compounds, 83 et seq.
Amino-group, replacement of diazo-group by, 52.
Aminonaphthols, diazotization of, 16.
Ammonia, action on diazo-compounds, 56.
Amyl diazoacetate, 106.
Amyl nitrite, use of, 6.
Aniline, 53.
Aurin, 30.
Azoammonium, 117.
Azobenzene, 82.
Azo-compounds, 80 et seq.
Azo-dyes, discovery of, 2, 4.
Azogen red, 26.
Azoimino-group, replacement of diazo-group by, 56, 58.
α-Azonaphthalene, 51.
Azophor blue D., 26.
Azophor red P.N., 26.
Azoxybenzene, 81.
Azoxy-compounds, 81.

Barium nitrite, use of, 10.
Benzeneazoacetaldehyde, 93.
Benzeneazoacetone, 92.
Benzeneazodiphenyl, 51, 60, 62.
Benzeneazomethane, 91.
Benzeneazonitroethane, 91.
Benzenediazoic acid, 102, 122, 155.
Benzenediazosulphonates, 128, 140, 155.
Benzidine, 23.
Benzonitrol, 26.
Benzoyl chloride, action on diazo-compounds, 59.
Bromobenzene, 45.
β-Bromonaphthalene, 46.

Calcium nitrite, use of, 9.
Chlorobenzene, 43.
1-Chlorodiazo-β-naphthalene nitrite, 69.
Chlorodibromodiazobenzene bromide, 68.
Cuprous chloride, rôle of, 44.
Cyanogen, replacement of diazo-group by, 49.
Cyano-group, replacement of diazo-group by, 49.

Decomposition of diazo-compounds, rate of, 35.

Diamines, 23.
Diazo, meaning of, 1, 117.
Diazoacetamide, 106.
Diazoacetic ester, 104.
Diazoacetophenone, 110.
Diazoaminobenzene, 73.
Diazoaminobenzoic acid, discovery of, 1.
Diazoamino-compounds, 73 et seq., 78.
Diazoaminomethane, 111.
Diazo-anhydrides, 148.
Diazo-azides, 9.
Diazobenzene chloride, 7, 17.
Diazobenzene hydroxide, 100, 120.
Diazobenzeneimide, 57, 58, 114.
Diazobenzene nitrate, 6, 11, 27, 28, 70, 116.
Diazobenzene picrate, 8.
Diazobenzene sulphate, 7, 27.
p-Diazobenzenesulphonic acid, 8, 144, 156.
Diazocamphor, 103.
Diazo-carbonates, 9.
Diazo-chromates, 8, 9.
Diazo-compounds, constitution of, 112 et seq.
Diazo-cyanides, 130, 140, 142, 143, 155.
Diazodiphenylamine, 153.
Diazo-ethers, 146.
Diazo-fluorides, 9.
Diazo-group, migration of, 76.
Diazo-halides, 138.
Diazo-hydroferricyanides, 9.
Diazo-hydroxides, 149, 150, 167.
Diazoic acids, 101.
Diazomethane, 108.
Diazomethanedisulphonic acid, 108.
Diazonaphthalenesulphonic acid, 8, 30.
Diazo-nitrates, 10.
Diazo-nitrites, 9.
Diazonium, 133.
Diazonium hydroxide, 150.
Diazo-oxides, 144, 156.
iso-Diazo-oxides, 100.
Diazo-perchlorates, 9.
Diazo-perhalides, 138, 139.
Diazophenols, 9, 11. See also Quinone-diazides.
p-Diazophenylhydroxylamine chloride, 11.
Diazo-picrates, 8.
Diazoprimuline, 70.
Diazo-salts, discovery of, 3.
Diazo-sulphides, 56. See also Thiodia-zoles.
Diazo-thiosulphates, 9.
Diazotization, 13 et seq., 28.
p-Diazotoluene nitrate, 11.
Diazo-tungstates, 9.

NAME INDEX

Oxford : Horace Hart, Printer to the University.

Mr. Edward Arnold's List of
Technical & Scientific Publications

Electrical Traction.

By ERNEST WILSON, Whit. Sch. M.I.E.E.,
Professor of Electrical Engineering in the Siemens Laboratory, King's College, London,
AND FRANCIS LYDALL, B.A., B.Sc.

NEW EDITION. REWRITTEN AND GREATLY ENLARGED.
Two volumes, sold separately. Demy 8vo., cloth.
Vol. I., with about 300 Illustrations and Index. 15s. net.
Vol. II., with about 170 Illustrations and Index. 15s. net.

In dealing with this ever-increasingly important subject the authors have divided the work into the two branches which are, for chronological and other reasons, most convenient, namely, the utilization of direct and alternating currents respectively. Direct current traction taking the first place, the first volume is devoted to electric tramways and direct-current electric railways. In the second volume the application of three-phase alternating currents to electric railway problems is considered in detail, and finally the latest developments in single-phase alternating current traction are discussed at length.

A Text-Book of Electrical Engineering.

By Dr. ADOLF THOMÄLEN.
Translated by GEORGE W. O. HOWE, M.Sc., Whit. Sch.,
A.M.I.E.E.,
Lecturer in Electrical Engineering at the Central Technical College, South Kensington.

With 454 Illustrations. Royal 8vo., cloth, 15s. net.

This translation of the " Kurze Lehrbuch der Electrotechnik " is intended to fill the gap which appears to exist between the elementary text-books and the specialized works on various branches of electrical engineering. It includes additional matter which is to be introduced into the third German edition, now in preparation. The book is concerned almost exclusively with principles, and does not enter into details of the practical construction of apparatus and machines, aiming rather at laying a thorough foundation which shall make the study of works on the design of machinery more profitable. Only the simplest elements of the higher mathematics are involved.

Alternating Currents.
A Text-Book for Students of Engineering.

By C. G. LAMB, M.A., B.Sc.,
Clare College, Cambridge ; Associate Member of the Institution of Electrical Engineers ;
Associate of the City and Guilds of London Institute.

viii + 325 pages. With upwards of 230 Illustrations. Demy 8vo., cloth,
10s. 6d. net.

The scope of this book is intended to be such as to cover approximately the range of reading in alternating current machinery and apparatus considered by the author as desirable for a student of general engineering in his last year—as, for example, a candidate for the Mechanical Sciences Tripos at Cambridge.

LONDON: EDWARD ARNOLD, 41 & 43 MADDOX STREET, W.

Electric and Magnetic Circuits.
By ELLIS H. CRAPPER, M.I.E.E.,
Head of the Electrical Engineering Department in the University College, Sheffield.
viii+380 pages.   Demy 8vo., cloth, 10s. 6d. net.

This, the introductory volume of a treatise on Electrical Engineering,
deals with the fundamental principles of Electricity and Magnetism,
and explains fully all the essential relationships of Electric and
Magnetic Circuits met with in continuous current working. It contains a very large number of worked examples, and several hundreds
of numerical examples taken from everyday practice.

Applied Electricity.
A Text-Book of Electrical Engineering for " Second Year" Students.
By J. PALEY YORKE,
Head of the Physics and Electrical Engineering Department at the London County Council
School of Engineering and Navigation, Poplar.
xii+420 pages.   Crown 8vo., cloth, 7s. 6d.

This volume is a text-book of Electrical Engineering for those who have
already become acquainted with the fundamental phenomena and
laws of Magnetism and Electricity.

Hydraulics.
By F. C. LEA, B.Sc., A.M.Inst.C.E.,
Senior Whitworth Scholar, A.R.C.S.; Lecturer in Applied Mechanics and Engineering Design,
City and Guilds of London Central Technical College, London.
With about 300 Illustrations.   Demy 8vo., 15s. net.

This book is intended to supply the want felt by students and teachers
alike for a text-book of Hydraulics to practically cover the syllabuses
of London and other Universities, and of the Institution of Civil
Engineers.

Hydraulics.
By RAYMOND BUSQUET,
Professeur à l'École Industrielle de Lyon.
AUTHORIZED ENGLISH EDITION.
Translated by A. H. PEAKE, M.A.,
Demonstrator in Mechanism and Applied Mechanics in the University of Cambridge.
viii+312 pages.   With 49 Illustrations.   Demy 8vo., cloth, 7s. 6d. net.

This work is a practical text-book of Applied Hydraulics, in which complete technical theories and all useful calculations for the erection of
hydraulic plant are presented.

Power Gas Producers.
Their Design and Application.
By PHILIP W. ROBSON,
Of the National Gas Engine Co., Ltd.; sometime Vice-Principal of the Municipal School of
Technology, Manchester.
Demy 8vo., cloth, 10s. 6d. net.

The recent enormous increase in the use of gas power is largely due to
the improvements in gas producers. This book, which is written
by a well-known expert, goes thoroughly into the theory, design,
and application of all kinds of plants, with chapters on working and
general management.

The Balancing of Engines.

By W. E. DALBY, M.A., B.Sc., M.Inst.C.E., M.I.M.E.,
Professor of Engineering, City and Guilds of London Central Technical College.

SECOND EDITION, REVISED AND ENLARGED.
xii + 283 pages. With upwards of 180 Illustrations.
Demy 8vo., cloth, 10s. 6d. net.

CONTENTS.

CHAP.
I. The Addition and Subtraction of Vector Quantities.
II. The Balancing of Revolving Masses.
III. The Balancing of Reciprocating Masses.—Long Connecting-rods.
IV. The Balancing of Locomotives.

CHAP.
V. Secondary Balancing.
VI. Estimation of the Primary and Secondary Unbalanced Forces and Couples.
VII. The Vibration of the Supports.
VIII. The Motion of the Connecting-rod.
APPENDIX.   EXERCISES.   INDEX.

Valves and Valve Gear Mechanisms.

By W. E. DALBY, M.A., B.Sc., M.Inst.C.E., M.I.M.E.,
Professor of Engineering, City and Guilds of London Central Technical College.

xviii + 366 pages. With upwards of 200 Illustrations.
Royal 8vo., cloth, 21s. net.

Valve gears are considered in this book from two points of view—namely, the analysis of what a given gear can do, and the design of a gear to effect a stated distribution of steam. The gears analyzed are for the most part those belonging to existing and well-known types of engines, and include, amongst others, a link motion of the Great Eastern Railway, the straight link motion of the London and North-Western Railway, the Walschaert gear of the Northern of France Railway, the Joy gear of the Lancashire and Yorkshire Railway, the Sulzer gear, the Meyer gear, etc. The needs of students and draughtsmen have been kept in view throughout.

" No such systematic and complete treatment of the subject has yet been obtainable in book form, and we doubt if it could have been much better done, or by a more competent authority. The language is exact and clear, the illustrations are admirably drawn and reproduced."—*The Times.*

The Strength and Elasticity of Structural Members.

By R. J. WOODS, M.E., M.Inst.C.E.,
Fellow and Assistant Professor of Engineering, Royal Indian Engineering College, Cooper's Hill.

SECOND EDITION, REVISED.
xii + 310 pages. With 292 Illustrations. Demy 8vo., cloth, 10s. 6d. net.

" To students for the final examination of the R.I.B.A. we can strongly recommend such a practical and thorough text-book."—*British Architect.*

" This is a practical book, and, although written mainly for engineering students, may be commended as one likely to prove equally useful to those engaged in active practice."—*Mechanical Engineer.*

Calculus for Engineers.

By JOHN PERRY, M.E., D.Sc., F.R.S.,
Professor of Mechanics and Mathematics in the Royal College of Science, London ; Vice-President of the Physical Society; Vice-President of the Institution of Electrical Engineers.

EIGHTH IMPRESSION.
viii + 382 pages. With 106 Illustrations. Crown 8vo., cloth, 7s. 6d.

Mathematical Drawing.

Including the Graphic Solution of Equations.

By G. M. MINCHIN, M.A., F.R.S.,

Professor of Applied Mathematics at the Royal Indian Engineering College, Cooper's Hill ;

AND JOHN BORTHWICK DALE, M.A.,

Assistant Professor of Mathematics at King's College, London.

Crown 8vo., cloth, 7s. 6d. net.

Graphic methods in Mathematics, which have attracted so much attention within the last few years, may be said to have attained a greatly increased importance by the decision of the University of London to require a knowledge of Mathematical Drawing from all candidates for the B.Sc. Degree.

The present work is largely an attempt to systematize somewhat vague methods of solving the non-algebraic equations which so often contain the solutions of physical problems.

Five-Figure Tables of Mathematical Functions.

Comprising Tables of Logarithms, Powers of Numbers, Trigonometric, Elliptic, and other Transcendental Functions.

By JOHN BORTHWICK DALE, M.A.,

Assistant Professor of Mathematics at King's College, London.

vi + 92 pages.   Demy 8vo., cloth, 3s. 6d. net.

This collection of Tables has been selected for use in the examinations of the University of London.

"This is a most valuable contribution to the literature of Mathematical refer-ence . . . . To anyone engaged in almost any form of higher physical research this compilation will be an enormous boon in the way of saving time and labour and collecting data. . . . The five-figure tables of roots and powers are, perhaps, the most useful features of the work."—*Mining Journal.*

Logarithmic and Trigonometric Tables (To

Five Places of Decimals). By JOHN BORTHWICK DALE, M.A., Assistant Professor of Mathematics at King's College, London. Demy 8vo., cloth, 2s. net.

Traverse Tables.

With an Introductory Chapter on Co-ordinate Surveying.

By HENRY LOUIS, M.A.,   AND   G. W. CAUNT, M.A.,

Professor of Mining and Lecturer on Surveying,   Lecturer in Mathematics,
                    Armstrong College, Newcastle-on-Tyne.

xxviii + 92 pages.   Demy 8vo., flexible cloth, rounded corners, 4s. 6d. net.

"The admirable, compact, and inexpensive tables compiled by Professor Henry Louis and Mr. G. W. Caunt. They are just what is required by the mining student and by the practical mine surveyor. . . . Their publication at a low price renders this convenient and rapid method of working out traverse surveys accessible to a class of workers from whom it has hitherto been debarred."—*Mining Journal.*

Organic Chemistry for Advanced Students.

By JULIUS B. COHEN, Ph.D., B.Sc.,
Professor of Organic Chemistry in the University of Leeds, and Associate of Owens College, Manchester.

Demy 8vo., cloth, 21s. net.

The book is written for students who have already completed an elementary course of Organic Chemistry, and is intended largely to take the place of the advanced text-book. For it has long been the opinion of the author that, when the principles of classification and synthesis and the properties of fundamental groups have been acquired, the object of the teacher should be, not to multiply facts of a similar kind, but rather to present to the student a broad and general outline of the more important branches of the subject. This method of treatment, whilst it avoids the dictionary arrangement which the text-book requires, leaves the writer the free disposal of his materials, so that he can bring together related substances, irrespective of their nature, and deal thoroughly with important theoretical questions which are often inadequately treated in the text-book.

The Chemical Synthesis of Vital Products and the Inter-relations between Organic Compounds.

By RAPHAEL MELDOLA, F.R.S., V.P.C.S., F.I.C., etc.,
Professor of Chemistry in the City and Guilds of London Technical College, Finsbury.

Vol. I., xvi + 338 pages. Super Royal 8vo., cloth, 21s. net.

The great achievements of modern Organic Chemistry in the domain of the synthesis or artificial production of compounds which are known to be formed as the result of the vital activities of plants and animals have not of late years been systematically recorded. The object of the present book is to set forth a statement, as complete as possible, of the existing state of knowledge in this most important branch of science. The treatment is calculated to make the volume a work of reference which will be found indispensable for teachers, students, and investigators, whether in the fields of pure Chemistry, of Chemical Physiology, or of Chemical Technology.

Lectures on Theoretical and Physical Chemistry.

By Dr. J. H. VAN 'T HOFF,
Professor of Chemistry at the University of Berlin.

Translated by R. A. LEHFELDT, D.Sc.,
Professor of Physics at the Transvaal Technical Institute, Johannesburg.

In three volumes, demy 8vo., cloth, 28s. net, or separately as follows :

Part I. CHEMICAL DYNAMICS. 254 pages, with 63 Illustrations. 12s. net.

Part II. CHEMICAL STATICS. 156 pages, with 33 Illustrations. 8s. 6d. net.

Part III. RELATIONS BETWEEN PROPERTIES AND COMPOSITION. 143 pages, 7s. 6d. net. '

Experimental Researches with the Electric Furnace.

By HENRI MOISSAN,
Membre de l'Institut ; Professor of Chemistry at the Sorbonne.

AUTHORIZED ENGLISH EDITION.

Translated by A. T. de MOUILPIED, M.Sc., Ph.D.,
Assistant Lecturer in Chemistry in the University of Liverpool.

xii + 307 pages, with Illustrations. Demy 8vo., cloth, 10s. 6d. net.

"There is hardly a page of it which is not crowded with interest, and hardly a section which does not teem with suggestion ; and if the coming of this English edition of the book has been so long delayed, we may still be thankful that it has come at last, and come in a form which it is a pleasure to handle and a delight to read."—*Electrical Review.*

Electrolytic Preparations.

Exercises for use in the Laboratory by Chemists and Electro-Chemists.

By DR. KARL ELBS,
Professor of Organic and Physical Chemistry at the University of Giessen.

Translated by R. S. HUTTON, M.Sc.,
Demonstrator and Lecturer on Electro-Chemistry at the University of Manchester.

xii + 100 pages. Demy 8vo., cloth, 4s. 6d. net.

The book contains a complete course of examples on the application of electrolysis to the preparation of both inorganic and organic substances. It will be found useful as filling a distinct gap in the text-book literature suitable for use in chemical laboratories, and should enable the chemist to make use of the many valuable and elegant methods of preparation which have been worked out during recent years, the advantages and ease of application of which he cannot appreciate without such a guide.

Introduction to Metallurgical Chemistry for Technical Students.

By J. H. STANSBIE, B.Sc. (LOND.), F.I.C.,
Associate of Mason University College, and Lecturer in the Birmingham University Technical School.

SECOND EDITION.

xii + 252 pages. Crown 8vo., cloth, 4s. 6d.

An Experimental Course of Chemistry for Agricultural Students.

By T. S. DYMOND, F.I.C., Lately Principal Lecturer in the Agricultural Department, County Technical Laboratories, Chelmsford. New Impression. 192 pages, with 50 Illustrations. Crown 8vo., cloth, 2s. 6d.

A History of Chemistry.

By DR. HUGO BAUER,
Royal Technical Institute, Stuttgart.

Translated by R. V. STANFORD, B.Sc. (LOND.),
Priestley Research Scholar in the University of Birmingham.

Crown 8vo., cloth, 3s. 6d. net.

The Becquerel Rays and the Properties of Radium.

By THE HON. R. J. STRUTT, F.R.S.,
Fellow of Trinity College, Cambridge.

SECOND EDITION, REVISED AND ENLARGED.

viii + 222 pages, with Diagrams.   Demy 8vo., cloth, 8s. 6d. net.

" If only a few more books of this type were written, there might be some hope of a general appreciation of the methods, aims, and results of science, which would go far to promote its study. . . . A book for which no praise can be excessive."— *Athenæum.*

Astronomical Discovery.

By HERBERT HALL TURNER, D.Sc., F.R.S.,
Savilian Professor of Astronomy in the University of Oxford.

xii + 225 pages, with Plates and Diagrams.   Demy 8vo., cloth, 10s. 6d. net.

An Introduction to the Theory of Optics.

By ARTHUR SCHUSTER, Ph.D., Sc.D., F.R.S.,
Recently Professor of Physics at the University of Manchester.

xvi + 340 pages, with Illustrations.   Demy 8vo., cloth, 15s. net.

" We know of no book written with a set purpose better adapted to serve the purpose for which it was written, nor any that the earnest student of optics will find more interesting and profitable. The work itself, without the confession of the preface, shows that Professor Schuster is a teacher, and every page bears evidence that he is a master of his subject . . . We heartily recommend the book to our readers."— *Ophthalmic Review.*

Wood.

A Manual of the Natural History and Industrial Applications of the Timbers of Commerce.

By G. S. BOULGER, F.L.S., F.G.S., A.S.I.,
Professor of Botany and Lecturer on Forestry in the City of London College, and formerly in the Royal Agricultural College.

NEW EDITION.   REVISED AND ENLARGED.   Demy 8vo., 12s. 6d. net.

" It is just the book that has long been wanted by land agents, foresters, and wood-men, and it should find a place in all technical school libraries."—*Field.*

Manual of Alcoholic Fermentation and the Allied Industries.

By CHARLES G. MATTHEWS, F.I.C., F.C.S., ETC.

xvi + 295 pages, with 8 plates, and 40 Illustrations.   Crown 8vo., cloth, 7s. 6d. net.

" This is a book worthy of its author, and well worth perusing by every student. . . . The student, both old and young, as well as the practical brewer, will find this book gives him some very useful information."—*Brewers' Guardian.*

The Evolution Theory. By DR. AUGUST WEIS-

MANN, Professor of Zoology in the University of Freiburg in Breisgau.
Translated, with the Author's co-operation, by J. ARTHUR THOMSON,
Regius Professor of Natural History in the University of Aberdeen ; and
MARGARET THOMSON. Two vols., xvi + 416 and viii + 396 pages, with over
130 Illustrations. Royal 8vo., cloth, 32s. net.

" The subject has never been so fully and comprehensively expounded before ; and
it is not necessary to subscribe to all the author's tenets in order to recognise the value
and the absorbing interest of his exposition, with its prodigious wealth of illustration,
its vast store of zoological knowledge, its ingenious interpretations and far-reaching
theories. English readers have reason to be grateful to Professor and Mrs. Thomson
for their admirable translation of a book which is not easy reading in any language,
but is indispensable for a thorough comprehension of the theory of evolution as it
stands to-day."—*The Times*.

The Chances of Death and Other Studies in

Evolution. By KARL PEARSON, M.A., F.R.S , Professor of Applied
Mathematics in University College, London, and formerly Fellow of King's
College, Cambridge. 2 vols., xii + 388 and 460 pages, with numerous
Illustrations. Demy 8vo., cloth, 25s. net.

Animal Behaviour. By Professor C. LLOYD

MORGAN, LL.D., F.R.S., Principal of University College, Bristol.
viii + 344 pages, with 26 Illustrations. Large crown 8vo., cloth, 10s. 6d.

This important contribution to the fascinating subject of animal psycho-
logy covers the whole ground from the behaviour of cells up to that
of the most highly developed animals.

" Professor Lloyd Morgan's book is of great value, not only for his own observations, but for the clear summary of what has been done by other workers in the same
field."—*Speaker*.

BY THE SAME AUTHOR.

Habit and Instinct. viii + 352 pages, with Photo-

gravure Frontispiece. Demy 8vo., cloth, 16s.

Professor ALFRED RUSSEL WALLACE :— " An admirable introduction to the study
of a most important and fascinating branch of biology, now for the first time based
upon a substantial foundation of carefully observed facts and logical induction from
them."

BY THE SAME AUTHOR.

The Springs of Conduct. Cheaper Edition.

viii + 317 pages. Large crown 8vo., cloth, 3s. 6d. This volume deals with
the Source and Limits of Knowledge, the Study of Nature, the Evolution of
Scientific Knowledge, Body, and Mind, Choice, Feeling, and Conduct.

BY THE SAME AUTHOR.

Psychology for Teachers. New Edition, entirely

rewritten. xii + 308 pages. Crown 8vo., cloth, 4s. 6d.

An Introduction to Child Study. By Dr. W. B.

DRUMMOND. Crown 8vo., cloth, 6s. net.

The Child's Mind : Its Growth and Training. By

W. E. URWICK, University of Leeds. Crown 8vo., cloth, 4s. 6d. net.

LONDON : EDWARD ARNOLD, 41 & 43 MADDOX STREET, W.